The Modulation of Dopaminergic Neurotransmission by Other Neurotransmitters

The Modulation of
Dopaminergic Neurotransmission
by Other Neurotransmitters

Edited by

Charles R. Ashby, Jr.

CRC Press

Boca Raton   New York   London   Tokyo

Library of Congress Cataloging-in-Publication Data

The modulation of dopaminergic neurotransmission by other
   neurotransmitters / edited by Charles R. Ashby, Jr.
         p.   cm.
   Includes bibliographical references and index.
   ISBN 0-8493-4780-7
   1. Dopamine.   2. Neurotransmitters.   3. Dopaminergic mechanisms.
I.. Ashby, Charles R.
   [DNLM: 1. Dopamine--physiology.   2. Neurotransmitters--physiology.
3. Receptors, Dopamine--physiology.   WK 725 M692 1995]
QP364.7.M636   1995
612.8'22--dc20
DNLM/DLC
for Library of Congress   95-43990
                                                                          CIP

© 1996 by CRC Press, Inc.

No claim to original U.S. Government works
International Standard Book Number 0-8493-4780-7
Printed in the United States of America 1 2 3 4 5 6 7 8 9 0
Printed on acid-free paper

THE EDITOR

Charles R. Ashby, Jr., Ph.D., is an Associate Scientist in the Medical Department at Brookhaven National Labs, Upton, New York.

Dr. Ashby obtained a B.A. degree in 1983 in psychology and biology at the University of Louisville, Louisville, Kentucky. In 1987, he was awarded a Ph.D. in pharmacology from the University of Louisville, School of Medicine. From 1987 to 1991, he was a postdoctoral fellow in the Department of Psychiatry at the State University of New York at Stony Brook and in 1991, he became a Research Assistant Professor in the Department of Psychiatry. He moved to Brookhaven National Labs in 1992 where he was appointed an Assistant Scientist. He was promoted to Associate Scientist in 1994. Dr. Ashby is now an Assistant Professor in the Department of Pharmaceutical Sciences at St. John's University.

Dr. Ashby is a member of the Society for Neuroscience, American Association for the Advancement of Science and the New York Academy of Sciences. He is the recipient of grants from the National Institute of Mental Health. He has authored 40 papers and 4 book chapters.

Dr. Ashby's research interests include studying the effects that manipulating 5-HT and various neuropeptides has on midbrain DA neurons in rats using *in vivo* electrophysiology. He is also interested in examining and using behavioral and electrophysiological methods to examine the dopaminergic system in strains of rats that display marked differences in their propensities to self-administer drugs of abuse.

CONTRIBUTORS

Charles R. Ashby, Jr., Ph.D.
Department of Pharmaceutical Sciences
St. John's University
Jamaica, NY 11439

Jonathan D. Brodie, M.D., Ph.D.
Department of Psychiatry
New York University
School of Medicine
New York, NY 10016

Louis A. Chiodo, Ph.D.
Department of Pharmacology
Texas Tech University
Health Sciences Center
3601 4th Street
Lubbock, TX 79430

Lynn Churchill, Ph.D.
Alcoholism and Drug Abuse Program
Washington State University
Pullman, WA 99164-6530

Stephen L. Dewey, Ph.D.
Chemistry Department
Bldg. 555
Brookhaven National Laboratory
Upton, NY 11973

Arthur S. Freeman, Ph.D.
Department of Pharmacology
Texas Tech University
Health Sciences Center
3601 4th Street
Lubbock, TX 79430

Peter W. Kalivas, Ph.D.
Alcoholism and Drug Abuse Program
Washington State University
Pullman, WA 99164-6530

Mark D. Kelland, Ph.D.
Department of Psychology
St. Anselm College
St. Anselm Drive
Manchester, NH 03102-1310

Michele L. Pucak, Ph.D.
Department of Psychiatry
University of Pittsburgh
3811 O'Hara St
W. 1650 BST
Pittsburgh, PA 15213

Gwen Smith, Ph.D.
PET Center
University of Pittsburgh
Pittsburgh, PA 15260

Jean-Pol Tassin, Ph.D.
INSERM U114
College de France
11, Place Marcelin-Berthelot
75231 Paris, Cedex 05, France

Judith R. Walters, Ph.D.
Experimental Therapeutics
National Institutes of Health
NINCDS
Building 10, Room 5C103
Bethesda, MD 20892

PREFACE

Research over the past four decades indicates that there are at least 80 potential neurotransmitters located in the brain. Indeed, there have been numerous studies examining the neuroanatomy, biochemistry, and pharmacology of these systems. In addition, a number of studies, particularly psychiatric and neurological ones, have focused on the one neurotransmitter-one disease concept. For example, it was originally hypothesized that schizophrenia was related to a dysregulation of dopamine (DA) neurons, whereas depression is the result of NE/5-HT neuronal dysregulation.

However, there is accumulating evidence indicating that neurotransmitters in the brain interact or modulate one another via complex afferent and efferent projections. Consequently, no neurotransmitter is an "island". Our knowledge regarding the interaction of neurotransmitters has increased greatly thanks to the synthesis and introduction of selective pharmacological agents for various receptors and uptake carriers and better neuroanatomical methods.

One of the most well characterized and highly researched neurotransmitter systems in the brain is DA. Since its discovery in the brain in 1959, numerous studies indicate that it plays an important role in translating motivationally relevant stimuli into adaptive motor responses. This has implications in psychiatric disorders such as schizophrenia and attention deficit disorder as well as Parkinson's disease. In addition, there are considerable data suggesting that DA neurons, particularly those in the mesotelencephalic system, play a role in mediating the rewarding/reinforcing effects of drugs of abuse. Thus, modulating DA function and/or activity could have significant physiological consequences.

It is known that DA cell bodies located in the ventral tegmental area (VTA) and substantia nigra pars compacta (SNC) are innervated by a number of neurotransmitters such as DA, GABA, enkephalins, cholecystokinin, and substance P, to name a few. The VTA and SNC also contain receptors for various neurotransmitters and neuromodulators. Furthermore, areas that receive dopaminergic innervation, such as the nucleus accumbens, amygdala, and striatum also receive innervation from other neurotransmitter systems. This is further complicated by the fact that these and other brain areas send feedback projections to DA cell bodies. Consequently, dopaminergic function can be modulated via the cell bodies (directly or indirectly) by direct synaptic contacts or terminal areas. The fine tuning of the DA system by other neurotransmitters could have important implications.

The purpose of this book is to present chapters that summarize or review behavioral, biochemical, radioligand binding, electrophysiological, and positron emission tomography evidence indicating that the function and activity of DA neurons are modulated by other neurotransmitter systems. The modulation of DA neurons by the peptides (enkephalins/endorphins and CCK) is reviewed in Chapters 2 and 4. The monoaminergic modulations (NE, 5-HT, and GABA) of the DA system are described in Chapters 1, 3 and 6. Finally, Chapter 5 will discuss the interaction of a number of neurotransmitters from the perspective of PET studies.

CONTENTS

Chapter 1

THE MODULATION OF DOPAMINERGIC NEUROTRANSMISSION BY NOREPINEPHRINE

Charles R. Ashby, Jr. and Jean-Pol Tassin

Chapter 2

DOPAMINE-OPIOID INTERACTIONS IN THE BASAL FOREBRAIN
Lynn Churchill and Peter W. Kalivas

Chapter 3

SEROTONERGIC MODULATION OF MIDBRAIN
DOPAMINE SYSTEMS

Mark D. Kelland and Louis A. Chiodo

Chapter 4

**ELECTROPHYSIOLOGICAL AND BIOCHEMICAL INTERACTIONS
BETWEEN DOPAMINE AND CHOLECYSTOKININ IN THE BRAIN**

Arthur S. Freeman

Chapter 5

THE MODULATION OF MIDBRAIN DOPAMINERGIC SYSTEMS BY GABA

Judith R. Walters and Michele L. Pucak

Chapter 6

A STRATEGY FOR MEASURING NEUROTRANSMITTER INTERACTIONS *IN VIVO* WITH POSITRON EMISSION TOMOGRAPHY (PET): NEUROPSYCHIATRY IMPLICATIONS

Gwen S. Smith, Stephen L. Dewey, and Jonathan D. Brodie

CHAPTER 1

THE MODULATION OF DOPAMINERGIC TRANSMISSION BY NOREPINEPHRINE

Charles R. Ashby, Jr. and Jean-Pol Tassin

I. INTRODUCTION

The locus coeruleus (LC), a structure located in the pontine area, contains the largest noradrenergic cell group in the brain. [Foote et al., 1983] Neuroanatomical studies have shown that noradrenergic neurons in the LC project to the mesencephalon, an area that contains DA cell bodies (A9 and A10 DA neurons), although it is unknown whether NE fibers make contact with DA cell bodies [Jones and Moore, 1977; Jones and Yang, 1985; Phillipson, 1979] Furthermore, areas that contain a high density of dopaminergic nerve terminals are also innervated by noradrenergic nerve terminals. [Berger et al., 1974; Lindvall et al., 1974; Thierry et al., 1973] Radioligand binding studies have also shown that brain areas which contain DA receptors also possess noradrenergic receptors. [Boyajian et al., 1987; Heat et al., 1985] Thus, it is conceivable that NE could modulate the activity or function of DA neurons.

The purpose of this article will be to review evidence indicating that the modulation of NE neurons alters the activity or function of DA neurons.

II. THE EFFECT OF NE AGONISTS AND ANTAGONISTS ON MIDBRAIN DA CELLS: ELECTROPHYSIOLOGICAL STUDIES

A series of studies by Svensson and Grenhoff and their colleagues have examined the effect of various adrenergic compounds on DA neurons in the substantia nigra pars compacta (SNc) and ventral tegmental area (VTA) of anesthetized male rats using in vivo extracellular single cell recording. The systemic administration of the α_2 receptor agonist clonidine regularizes the firing pattern of DA cells in the VTA and the SNc [Grenhoff and Svensson, 1988, 1989]. This regularization was represented by a decrease in the variation coefficient without altering the firing rate of the cells. This effect was significantly attenuated by pretreating animals with idazoxan (0.5 mg/kg, i.v.) or yohimbine (1 mg/kg, i.v.) but not phentolamine (1 mg/kg, i.v.), an α_2 antagonist that does not produce central α_2 antagonism. The neuromodulatory effect of clonidine was abolished by pretreating animals with reserpine.

0-8493-4780-7/96/$0.00+$.50
© 1996 by CRC Press, Inc.

These results suggest that the modulation of VTA and SNc DA cell activity by clonidine is indirect and dependent on endogenous monoamines. Clonidine's action is probably not related to its alterationin blood pressure since although phentolamine produces a significant decrease in arterial blood pressure, it does not alter VTA DA cell activity. In addition, the systemic administration of idazoxan and yohimbine produced actions that were opposite to that observed for clonidine on DA neuronal activity. It is unlikely that clonidine is exerting its action by acting directly on α_2 receptors in the SNc and VTA as it has been reported that neither SNc nor VTA areas contain mRNA for α_{2A}, α_{2B} or α_{2C} receptors [Scheinin et al., 1994] and the density of these receptors in the two areas is low.

Grenhoff et al. (1993) examined the effect of the electrical stimulation of the LC on midbrain DA neurons in anesthetized rats. The regular, single pulse LC stimulation (0.5 Hz) produced an excitation (period of enhanced activity) followed by an inhibition (period of decreased activity) of midbrain DA neurons. The stimulation of the LC in 20 Hz bursts produced a longer period of inhibition only with repeated stimulation. The systemic administration of prazosin (0.6 mg/kg, i.v.) significantly attenuated the excitatory response following LC stimulation but had no effect on the inhibitory response. Overall, neither idazoxan or timolol altered the inhibition or excitation elicited by LC stimulation. The pretreatment of rats with reserpine abolished the excitatory and inhibitory responses. Overall, the results indicate that the LC-induced excitation of midbrain DA neurons is related to the release of NE which subsequently interacts with α_1 receptors.

It has been reported that the administration of prazosin (0.15-0.6 mg/kg, i.v.) produces a dose-dependent decrease in the burst firing of VTA DA cells without altering the firing rate. [Grenhoff and Svensson, 1993] The pretreatment of animals with reserpine abolished prazosin's action. These findings are consistent with the aforementioned results suggesting that the α_1 receptor plays an important role in mediating the activity of DA neurons.

A recent study has shown that the activity of DA cells in the VTA and SNc may be differentially regulated by α_1 receptors. [Andersson et al., 1994] The pretreatment of animals with prazosin (0.3 mg/kg, i.v.) significantly potentiates the increase in the basal firing rate of VTA DA cells elicited by 20-160 μg/kg i.v. of raclopride, a D_2/D_3 receptor antagonist. However, the raclopride-induced burst firing of VTA, but not that of SNc DA cells, was significantly decreased by prazosin. Furthermore, the pretreatment of animals with prazosin significantly attenuated the increase in extracellular DA levels in the nucleus accumbens, a major target area of VTA DA neurons, produced by a high dose (2560 μg/kg, i.v.) but not by a low dose (80 μg/kg, i.v.) of raclopride. In contrast, the increase in striatal DA levels produced by 80 and 2560 μg/kg of raclopride was not altered by pretreatment with prazosin. These results suggest that α_1 receptors may play an important role in modulating the activity of DA cells in the VTA but not the SNc. The authors suggested that the modulatory effects of prazosin on mesolimbic DA activity may occur at the cortical level and is mediated via a polysynaptic corticofugal pathway. It will be of considerable interest to determine the effect of compounds that are selective for the various subtypes of the α_1 receptor on DA neuronal activity.

Overall, the electrophysiological studies suggest that pharmacological manipulations that increase NE neurotransmission (i.e. LC stimulation, idazoxan administration) produce an increase in the irregularity of midbrain DA neurons. In contrast, experimental manipulations or treatments that decrease NE neurotransmission appeared to induce DA neurons to fire in a more regular pattern (i.e. decrease in irregularity). It is known that DA neurons firing in an irregular or burst firing pattern induce a greater release of DA from DA nerve terminals compared to DA neurons firing in a regular pattern or mode. [Gonon, 1988] Therefore, it may be postulated that the facilitation and attenuation of NE transmission will augment and decrease DA neurotransmission, respectively. Additional studies must be conducted in order to ascertain at what site (s) various NE compounds interact to produce their effects on DA neurons.

III. THE EFFECT OF EXPERIMENTAL MANIPULATION OF NE ON THE DA LEVELS IN THE BRAIN.

A. Dopamine-ß-Hydroxylase Inhibitors: It has been shown that the administration of DBH inhibitors such as FLA-63, U-14,624, disulfiram and diethyldithiocarbamate produce a significant, but not complete, depletion of NE without lesioning NE neurons. [Corrodi et al., 1970; Thornburg and Moore, 1971] Therefore, such compounds can produce NE depletion without inducing neurotoxicity. A number of studies have examined the effect of NE depletion by DBH inhibitors on brain levels of DA and the results are summarized in Table 1. Overall, the results are inconsistent, with some studies indicating that DBH inhibitors produce an increase or no change in brain DA levels. One problem with regards to the interpretation of the results is that it has been shown that several DBH inhibitors can produce peritoneal irritation upon injection which can induce stress in animals. [Thornburg and Moore, 1971] In addition, it has been reported that FLA-63 and U-14,624 have proconvulsant activity. [McKenzie and Soroko, 1973] Thus, it is possible that DA levels could be altered by stress and not by

Table 1. The effect of the depletion of NE by dopamine-β-hydroxylase inhibitors on brain DA levels in rats and mice: postmortem studies

Reference	Brain area	Compound	Effect on DA	Species
1	whole brain	500 mg/kg, s.c. DDC	0	mice
2	whole brain	3 x 50 mg/kg, DDC 3 x 100 mg/kg, DDC	↑ ↑	mice
3	whole brain	1 x 50 mg/kg, DDC 3 x 50 mg/kg, DDC	0 ↑	rats
	whole brain	0.02 - 20 mg, DDC i.c.v.	↑[a]	rats
4	whole brain	50 mg/kg, i.p., disulfiram	0	mice
5	whole brain	100 mg/kg, i.p.,disulfiram	0	mice
6	whole brain caudate brain stem	400 mg/kg,i.p.,disulfiram 400 mg/kg,i.p.,disulfiram 400 mg/kg,i.p.,disulfiram	0 0 ↑	rats rats rats
7	whole brain	50 mg/kg, FLA-63	0	mice
8	whole brain	5–80 mg/kg, FLA-63	↑[b]	mice
9	whole brain striatum limbic system hemispheres brain stem	30 mg/kg, i.p. FLA-63	↑ 0 ↑ ↑ ↑	rat
10	whole brain	25 mg/kg, FLA-63	0	rats

Table 1. (cont'd) The effect of the depletion of NE by
dopamine-ß-hydroxylase inhibitors on brain DA levels in
rats and mice: postmortem studies

Reference	Brain area	Compound	Effect on DA	Species
11	whole brain	25 mg/kg, U, 14,624	↑[c]	mice
12	whole brain	12.5–200 mg/kg, U, 14,624	↑[c]	mice
	whole brain	0.1–1% U, 14,624 in diet	0	mice

[a] An increase in DA levels was produced by all amounts of DDC; [b] Increases were produced by 40 and 80 mg/kg;
[c] Only at the 200 mg/kg dose; DDC, diethyldithiocarbamate

References: 1, Carlsson et al., 1966; 2, Maj et al., 1967; 3, Kleinrok et al., 1970; 4, Svensson and Waldeck, 1969;
5, Thornburg and Moore, 1971; 6, Goldstein and Nakajima, 1967; 7 Svensson, 1970; 8, Svensson, 1973; 9, Anden
et al., 1973; 10, Al-Shabibi and Doggett, 1978; 11, Johnson et al., 1971; 12, Von Voigtlander and Moore, 1973.

NE depletion. Furthermore, the increase in brain DA levels produced by some of the DBH inhibitors occurred only at high doses that did not inhibit DBH to a greater degree than lower doses, suggesting that this effect is not related to DBH inhibition. Finally, it is difficult to directly compare the results as there were differences between studies regarding the species of animals used and the brain areas examined.

B.   Lesioning of NE Neurons: The effect of experimental manipulations that destroy NE neurons is shown in Table 2. The results from the transection studies are inconclusive, with studies showing an increase, decrease or no change in specific brain areas. It is possible that this may be related to differences in the type of transection (unilateral vs. bilateral), the degree of NE depletion obtained and the length of time after the lesion.

Two recent microdialysis studies have shown that the lesioning of NE neurons by the selective NE neurotoxin DSP-4 [Jonsson et al., 1981] or by bilateral administration of 6-OHDA into the LC produce a decrease in basal extracellular levels of DA in the caudate and nucleus accumbens. [Lategan et al., 1990, 1992] The consistency of these results may be related to the fact that the experiments were conducted using microdialysis, which allows for the collection of samples in vivo and thereby removes problems related to the determination of DA levels in vitro. In addition, the degree of NE depletion was similar in both studies (75% vs. 67%).

Interestingly, the destruction of LC neurons has been shown to significantly alter the recovery of DA levels after the lesioning of DA neurons. Maviridis et al. [1991] examined the effect of lesioning the locus coeruleus on the loss of SNc cells and DA levels in the caudate and putamen of squirrel monkeys given 1-methyl-4-phenyl-1,2,3,6-tetrahydropyridine (MPTP), which destroys SNc DA cells, depletes striatal DA levels and induces parkinsonian symptoms in primates. In sham lesion controls, MPTP produced a 45% decrease in putamen DA levels and a moderate loss of cell bodies in the SNc at 9 weeks after MPTP treatment. In contrast, the LC lesioned monkeys exhibited a profound decrease in DA content in the caudate (84%) and putamen (91%) and severe neuronal cell loss in the SNc. Therefore, it seems that intact NE neurons are required to produce some degree of recovery of DA function.

IV. THE EFFECT OF NE AGONISTS AND ANTAGONISTS ON DA RELEASE FROM BRAIN SLICES AND ON EXTRACELULLAR LEVELS OF DA AS DETERMINED USING MICRODIALYSIS.

The effect of various adrenergic agonists and antagonists on DA release in brain slice preparations has been examined (Table 3).

Table 2. The effect of adrenergic neuronal lesioning on DA levels in the rat brain

Reference	Technique	Experimental methodology	Brain area	Effect on DA	Species
1	post-mortem	bilateral ventral NE bundle transection	NAc lateral septum nucleus tractus diagonalis	↑ ↑ ↑	rat
	post-mortem	unilateral ventral NE bundle transection	caudate olfactory tubercle NAc Stria terminalis VTA substantia nigra reticulata	↓ ↓ ↓ ↓ ↑ ↑	rat
2	post-mortem	MPTP + bilateral LC lesion with 6-OHDA[a]	striatum	↓	squirrel monkey
3	microdialysis[b]	bilateral LC lesion with 6-OHDA	NAc caudate	↓ ↓[c]	rat
4	microdialysis[b]	50 mg/kg i.p. DSP-4	NAc caudate	↓ ↓[c]	rat
5	post-mortem	Dorsal/NE bundle lesion	striatum ipsilateral contralateral	0 0	rat

[a] Animals were given; [b] Experiments were conducted in anesthetized rats; [c] The decrease in extracellular DA levels was greater in the caudate.

References: 1, O'Donohue et al., 1979; 2, Mavridis et al., 1991; 3, Lategan et al., 1990; 4, Lategan et al., 1992; 5, Reches and Meiner, 1992.

Table 3. The effect of adrenergic agonists and antagonists on DA release in the brain

Reference	Experimental preparation	Brain area	Drug concentration	Effect on DA release	Species
1	in vitro, slice	striatum	1 μm (−)-isoproterenol 1 μm (+)-isoproterenol	↑ 0	rats
2	in vitro, slice	hypothalamus	0.1 μm adrenaline	↑	rats
3	in vitro, slice	NAC	10 μm clonidine 1 μm isoproterenol 10 μm	↓ ↑ ↑	rats
4	in vitro, slice	striatum	0.1–2 μm UK–14304[a]	↓	rats
5	push–pull cannula	caudate	1 μm (−)-isoproterenol	↑	cat

[a] UK–14304 is an α_2–agonist.

References: 1, Reisine et al., 1982; 2, Ueda et al., 1983; 3, Nurse et al., 1985; 4, Russell et al., 1993; 5, Reisine et al., 1982.

Overall, the studies indicate that the incubation of brain slices from the striatum, nucleus accumbens and hypothalamus with the ß-adrenergic agonist (-)-isoproterenol produces an increase in the release of ^{3}H-DA and this can be blocked by the non-selective ß-adrenergic antagonist propranolol. In contrast, the α_2 receptor agonists UK 14304 and clonidine produced a decrease in the basal release of ^{3}H-DA from caudate slices. Similarly, the incubation of hypothalamic slices with yohimbine and adrenaline also decreased ^{3}H-DA release.

Recent studies have employed the technique of in vivo microdialysis to examine the effect of noradrenergic agents on extracellular DA levels in various brain areas. The systemic administration of oxaproptiline and desipramine have been shown to increase extracellular DA levels in the prefrontal cortex, but not in the dorsal striatum of rats. [Carboni et al., 1990] The increase in extracellular DA levels elicited by haloperidol in the prefrontal cortex, but not in the dorsal striatum, was potentiated by both oxaproptiline and desipramine. The lesioning of the dorsal NE bundle by the administration of 6-OHDA into the superior cerebellar peduncle prevented the ability of oxaproptiline to increase extracellular DA levels in the prefrontal cortex. These results suggest that DA can be removed from the extracellular space by reuptake into NE terminals in the prefrontal cortex, an area with a high density of NE.

A recent study examined the effect of various monoamine uptake blockers on DA (NE and 5-HT were also measured) levels in the VTA after their administration via the dialysis probe. [Chen and Reith, 1994] Overall, they reported that the efficacy of a compound to increase dialysate output of DA in the VTA of rats is positively correlated (r = 0.97 at EC_{100}) to its ability to increase dialysate NE levels. The authors state that the NE system contributes to DA release in the VTA, presumably through both transporter- and receptor-mediated mechanisms. It is possible that the released DA could be taken back up into NE nerve terminals and subsequently co-released with NE. This scenario is supported by data indicating that desipramine, a selective NE uptake blocker, decreases ^{3}H-DA accumulation in substantia nigra slices and increases DA levels in the rat prefrontal cortex. [Carboni et al., 1990; Kelley et al., 1985] In addition, this is consistent with evidence indicating the presence of NE nerve terminals within the VTA. [Gehlert et al., 1993; Javitch et al., 1985] This apparent facilitation of NE on DA neurotransmission is consistent with the aforementioned microdialysis data indicating that NE depletion produces a decrease in basal extracellular DA levels in the striatum and nucleus accumbens and with data indicating that the MPTP-induced decrease in DA levels is potentiated in locus-coeruleus lesioned animals. [Lategan et al., 1990, 1992; Mavridis et al., 1992]

Furthermore, this finding is in parallel with the previously mentioned electrophysiological evidence indicating that the facilitation of NE activity produces an increase in the irregularity of DA neuronal firing, which could ultimately produce an increase in terminal DA release. [Grenhoff et al., 1993; Svensson and Grenhoff, 1993]

V. THE EFFECT OF THE ACUTE AND CHRONIC ADMINISTRATION OF ADRENERGIC COMPOUNDS ON DA UTILIZATION, SYNTHESIS AND TURNOVER: POST MORTEM STUDIES

The initial biochemical studies that examined the effect of NE agonists and antagonists on DA function measured DA utilization and synthesis postmortem.   The utilization of DA was assessed by measuring the disappearance of DA in the brain one hour after the administration of D,L-α-methyl-p-tyrosine (AMPT), an inhibitor of tyrosine hydroxylase, the rate-limiting enzyme in the synthesis of DA. The synthesis of DA was determined by measuring the accumulation of L-DOPA, the immediate precursor of DA, 30 minutes following the injection of the aromatic amino acid decarboxylase inhibitor NSD-1015. Subsequent studies measured the level of DA and its primary metabolites, HVA and DOPAC, using the technique of high pressure liquid chromatography.   The results of these studies are summarized in Tables 4 and 5.

It has been reported that the administration of the α_2 agonist clonidine and the non-selective α_2 antagonist phenoxybenzamine produces a decrease in DA utilization, synthesis or turnover. [Anden et al., 1970; Anden and Grawboska, 1976; Anden and Strombom, 1974; Geyer and Lee, 1984; van Oene et al., 1984]   In contrast, the α_2 antagonists piperoxan, SKF 64139, SKF 72223 and yohimbine increases DA synthesis in the striatum and whole brain. [Anden et al., 1976; Anden et al., 1982; Anden and Grawboska, 1976; van Oene et al., 1984; Pettibone et al., 1985; Rabey et al., 1981; Strombom, 1976]   Both piperoxan and rauwolscine increase HVA and DOPAC levels and DA utilization in the striatum. [Geyer et al., 1984; Scatton et al., 1983; Waldmeier et al., 1982]   Similarly, yohimbine increase DA utilization of whole brain DA in mice and this effect is antagonized by prazosin. [Anden et al., 1982]   In contrast to the above studies, Rochette and Bralet [1974] reported that 50 μg/kg i.p. of clonidine did not alter DA synthesis in the striatum, but significantly decreased DA synthesis in the telencephalon.   Persson [1970] has shown that the systemic administration of 1-3 mg/kg of clonidine increased DA synthesis in the brain. It is possible that the difference between the results obtained by Persson and Rochette and Bralet may be related to the dose of

Table 4. The effect of adrenergic receptor agonists and antagonists on L-DOPA synthesis and α-methyl-p-tyrosine (AMPT)-induced disappearance of DA in the rat brain

References	Brain area	Adrenergic drug and dose (mg/kg)	Effect on L-DOPA synthesis	Effect on DA Disappearance
1	whole brain	20mg/kg, yohimbine	—	↑
2	striatum	3 mg/kg, yohimbine	↑	—
		20 mg/kg, phenoxybenzamine	↑	— 0[b]
3	striatum	5–20mg/kg, RX 781094[a]	—	0[b]
4	striatum	10mg/kg, idazoxan	—	○○
		yohimbine 1	—	←
		3	—	←
		10	—	→
		rauwolscine 1	—	→
		3	—	→
		10	—	→
		piperoxan 10	—	→
		20	—	
5	striatum	clonidine 0.1	→	—
		0.3	→	—
		1	→	—
		3	0	—
		yohimbine 3	←	—
		10	←	—
		phenoxybenzamine 20	0	—
6	striatum	25mg/kg, SKF 64139[c]	↑	—
		25mg/kg, SKF 72223[c]	↑	—

Table 4. (cont'd) The effect of adrenergic receptor agonists and antagonists on L-DOPA synethesis and α-methyl-p-tyrosine (AMPT)-induced disappearance of DA in the rat brain

References	Brain area	Adrenergic drug and dose (mg/kg)	Effects on L-DOPA synthesis	Effect on DA disappearance
7	striatum telencephalon	0.05mg/kg clonidine	— —	0 ↓

[a]Rx781094 = idazoxan; [b]no effect at any dose; [c]α₂ receptor antagonists

References: 1, Papeschi and Theiss, 1975; 2, Anden and Grawbowska, 1976; 3, Flockhart et al., 1982; 4, Scaffon et al., 1983; 5, Anden and Grawbowska, 1976; 6, Rabey et al., 1981; 7, Rochete and Bralet, 1974.

Table 5. The effect of NE receptor agonists and antagonists on
DA, HVA and DOPAC levels in the brain

Reference	Brain area	Compound given	Dose (mg/kg)	DA	DOPAC	HVA	Species
1	striatum	clonidine	0.05	→o	—	—	rats
	talencephalon	clonidine		o			
2	whole brain	clonidine	0	oo	—	—	rats
		phenoxybenzamine	20	oo	—	—	
		yohimbine	10	oo	—	—	
3	striatum	corynanthine	10	—	o	→	rats
			20	—	←o	o	
		piperoxan	1	—	o	o	
			3	—	o	←	
			10	—	o	o	
		prazosin	1	—	o	o	
			3	—	o	o	
			10	—	o	←	
		rauwolscine	1	—	o	←	
			2	—	o	o	
			5	—	o	o	
		tolazoline	25	—	o	o	
			50	—	←o	←	
	mesolimbic	corynanthine	10	—	o	←	
			20	—	o	o	
		piperoxan	1	—	o	o	
			3	—	←o	←	
			10	—	o	o	
		prazosin	1	—	o	→	
			3	—	o	o	
			10	—	o	→o	
		rauwolscine	1	—	o	→	
			2	—	←o	←	
			5	—	←	←	

Table 5. (cont'd) The effect of NE receptor agonists and antagonists on DA, HVA and DOPAC levels in the brain

Reference	Brain area	Compound given	Dose (mg/kg)	DA	DOPAC	HVA	Species
4	striatum	prazosin	2μmols/kg	0	0	↓	rats
		yohimbine	5μmols/kg	0	↑	↑	
		phenylephrine	25μmols/kg	0	0	↓	
		tramatoline	25μmol/kg	0	0	↓	
		DPI	25μmol/kg	0	0	↓	
5	striatum	idazoxan	1	0	0	0	rats
			3	↑	0	0	
			10	0	0	0	
			20	—	0	0	
		rauwolscine	1	0	↑	0	
			3	0	↑	↑	
			10	0	↑	↑	
		piperoxan	2.5	0	0	0	
			10	0	↑	↑	
			20	—	↑	↑	
			60	—	↑	↑	

Table 5. (cont'd) The effect of NE receptor agonists and antagonists
on DA, HVA and DOPAC levels in the brain

Reference	Brain Area	Compound given	Dose (mg/kg)	DA	DOPAC	HVA	Species
5	striatum	yohimbine	1	—	0	—	rats
			3	—	0	—	
			10	—	↑	—	
		prazosin	3	—	↓	—	
			10	—	↓	—	
			20	—	↓	—	
		phentolamine	10	—	0	—	
			20	—	↓	—	
			30	—	0	—	
6	striatum	clonidine	0.05	0	—	↑	rats
		yohimbine	10	0	—	↓	
7	striatum	prazosin	10	0	↓	↓	rats
		phenoxybenzamine	10	0	0	↓	
		phentolamine	10	0	0	0	
		piperoxan	10	0	↑	↑	
		yohimbine	10	0	↑	↑	
		propranolol	10	0	0	0	
		tolazoline	25	0	0	0	

Table 5. (cont'd) The effect of NE receptor agonists and antagonists on DA, HVA and DOPAC levels in the brain

Reference	Brain area	Compound given	Dose (mg/kg)	DA	DOPAC	HVA	Species
7	olfactory tubercle	prazosin	10	0	↓	↓	rats
		yohimbine	10	0	↑	↑	
8	striatal	rauwolscine	3.6	—	↑	↑	rats
		RX 781094	1–20	—	0	0	
		piperoxan	60	—	—	↑	
9	whole brain	phenoxybenzamine	2.5	0	—	—	rats
			5	0	—	—	
			10	0	—	—	
		phentolamine	1	0	—	—	rats
			5	0	—	—	
			10	0	—	—	

References: 1, Rochette and Bralet, 1975; 2, Kehr, 1981; 3, Waldmeier et al., 1982; 4, Van Oene et al., 1982; 5, Scatton et al., 1983; 6, Geyer and Lee, 1984; 7, Van Oene et al., 1984; 8, Dedek et al., 1982; 9, Andes and Strombom, 1974.

clonidine used, as at low doses, clonidine stimulates NE receptors, but at high doses, it is adrenolytic.

The α_1 receptor antagonist corynanthine increases, whereas another α_1 receptor antagonist prazosin has been shown to produce either no significant effect or a decrease in DA turnover in various brain areas. [van Oene et al., 1984; Scatton et al., 1980; Waldmeier et al., 1982] It has been shown that prazosin increases whole brain DA levels after AMPT administration in mice. [Anden et al., 1982]

Based on the above results, it may be generally concluded that α_1 receptor antagonists decrease, whereas α_2 antagonists increase DA utilization and turnover. However, this may not be completely true as it has been shown that compounds such as yohimbine, rauwolscine, corynanthine and piperoxan (high dose) possess direct antidopaminergic activity. [van Oene et al., 1984; Scatton et al., 1980; Waldmeier et al., 1982] In addition, the administration of the selective α_2 receptor antagonist idazoxan does not produce a significant alteration in DA synthesis. [Chapleo et al., 1983; Pettibone et al., 1985; Scatton et al., 1983] There is now a significant amount of evidence indicating that there are subtypes of the α_1 and α_2 receptors in the brain. [Bylund et al., 1988, 1994] Therefore, the interpretation of some of the aforementioned results is hampered by this fact as none of the compounds used in the studies show selectivity. Furthermore, the use of different doses, routes of administration and animal species also complicates the interpretation of the data and comparisons of the individual studies.

VI. THE EFFECT OF NE AGENTS ON DA MEDIATED BEHAVIORS

A. Amphetamine and Apomorphine-Induced Stereotypy and Locomotor Activity: Role of DA Neurons

It has been shown that low doses of amphetamine produces behavioral hyperactivity (increased locomotor activity) whereas as high doses elicit stereotypy (invariant, repetitive behaviors). In contrast, low doses of apomorphine induce behavioral hypoactivity, although high doses elicit locomotor hyperactivity and stereotypy. There is considerable evidence indicating that the behavioral effects elicited by amphetamine and apomorphine are predominantly mediated by DA neurons. [Clarke et al., 1988; Costall et al., 1977; Creese and Iversen, 1975; Hollister et al., 1974; Jackson et al., 1975; Kelley et al., 1975; Makanjuola et al., 1980; Pinjenberg et al., 1975] Specifically, the alterations in locomotor activity appear to be mediated by mesolimbic structures, whereas stereotypy behaviors are more closely associated with the nigrostriatal system in rats. [Costall et al., 1977; Kelley et al., 1975] The administration of various DA antagonists effectively blocks amphetamine and apomorphine-induced behaviors, particularly locomotor hyperactivity and stereotypy. In contrast, the administration of various α and ß-adrenergic receptor antagonists do not significantly alter d-amphetamine or apomorphine-induced locomotor activation. [for review, see Kokkinidis and Anisman, 1980; Fishman et al., 1983]

Previously, it had been suggested that NE neurons played a primary role in mediating the behavioral hyperactivity elicited by amphetamine. [Corrodi et al., 1970; Randrup and Scheel-Kruger, 1966; Svensson, 1970; Svensson and Waldeck, 1969] However, as stated above, the behavioral effects of amphetamine appear to be mediated by DA neurons. Nonetheless, there is evidence indicating that NE plays an important role in modulating the behavioral effect produced by dopaminergic agents (see Table 6).

1. Adrenergic Agonists

The results in Tables 6 and 7 appear to indicate that the pretreatment of animals with the α receptor agonists potentiates certain stereotyped behaviors induced by amphetamine and apomorphine. [Mogilnicka and Braestrup, 1976; Muller and Nyhan, 1982; Thomas and Handley, 1978; Zetler, 1985; however, see Pfil and Hornykiewicz, 1985 and Pycock et al., 1977] Luttinger and Durivage [1986] have reported that amphetamine-induced locomotor activity in mice is not antagonized by clonidine. However, the effect of adrenergic agonists on the locomotor activity elicited by apomorphine and amphetamine have been mixed. [Dolphin et al., 1977; Mueller and Nyhan, 1982; Strombom, 1976]

2. NE Synthesis Inhibitors

The effect of NE synthesis inhibitors on the behaviors induced by amphetamine and apomorphine is equivocal (Table 8). For example, the dopamine-ß-hydroxylase inhibitors (DBH) diethyldithiocarbamate (DDC) and FLA-63 potentiate amphetamine's action. [Grabowska-Anden, 1977; Mogilnicka and Braestrup, 1976] However, other studies using DDC and the more selective DBH inhibitors U-14,624 and FLA-63 have shown that amphetamine's action is either slightly decreased or unaffected. [Carlsson, 1970; Corrodi, 1970; Hollister et al., 1974; Johnson, 1970; Ljungberg and Ungerstedt, 1976; Rolinski and Scheel-Kruger, 1973; Svensson, 1970; Thornburg and Moore, 1973] Similarly, it has been reported that FLA-36 has no significant effect on apomorphine-induced ambulation, whereas apomorphine-induced stereotypy is potentiated by FLA-63. [Ljungberg and Ungerstedt, 1976; Mogilnicka and Braestrup, 1976]

Table 6. The effect of adrenergic agonists on amphetamine-
and apomorphine-induced stereotypies in rats and mice

Reference	Challenge drug and dose (mg/kg)	Adrenergic agonists and dose (mg/kg)	Effect on behavior	Species
1	apomorphine, 0.5, i.p.	clonidine 0.25	↑	rats
2	apomorphine, 0.01 mg. (intra-NAC)	clonidine 0.05	↑	rats
3	apomorphine, 0.1-5, s.c.	clonidine 0.5, i.p.	0[a]	rats
4	amphetamine, 6, s.c.	clonidine 0.01 0.05 0.50	↑ ↑ ↑	rats
5	amphetamine, 7, i.p.	St-587	0[b]	rats
6	apomorphine, 12.5, s.c.	clonidine 0.19-1.2	↑	mice
7	amphetamine, i.p. 0.25 0.50 1	modafinil 32 64 128	0 0 0	rats

[a] Clonidine did not significantly alter the stereotypies induced by any dose of apomorphine.
[b] Dose of St-587 was not stated in the text.

References: 1, Mogilnicka and Braestrup, 1976; 2, Dolphin et al., 1977; 3, Pycock et al.,1977; 4, Muller and Nyhan, 1982; 5, Pfil and Hornykiewicz, 1985; 6, Zetler, 1985; 7, Dutell et al., 1990.

Table 7. The effect of adrenergic agonists on apomorphine- and amphetamine-induced locomotor activity in rats and mice

Reference	Challenge drug and dose (mg/kg)	Adrenergic agonist and dose (mg/kg)	Effect on behavior	Species
1	apomorphine, 1, s.c.	clonidine 0.2 1 5 25	0 0 ↑ ↑	mice
2	apomorphine, 3, i.p.	clonidine 1.5 3	↑ ↑	mice
3	apomorphine, 0.05 0.2 0.8	clonidine 0.025–0.8	↓[a] ↓[b] ↓[c]	rats
4	amphetamine, 2, s.c.	clonidine 0.01 0.05 0.50	↑ ↑ ↑	rats
5	amphetamine, 1, i.p.	St-587	↑	rats

[a] decrease at 0.1 and 0.2 mg/kg; [b] decrease at 0.05, 0.1, 0.2 and 0.8 mg/kg; [c] decrease at 0.8 mg/kg.

References: 1, Maj et al., 1972; 2, Anden and Strombom, 1974; 3, Strombom, 1976; 4, Muller and Nyhan, 1982; 5, Pfil and Hornykiewicz, 1985.

Table 8. The effect of dopamine ß-hydroxylase inhibitors on
behaviors elicited by d-amphetamine administration in rats and mice

Reference	Behavior assessed	Dose of amphetamine (mg/kg)	DBH inhibitor and dose (mg/kg)	Effect on behavior
1	locomotor activity	5, s.c.	DDC, 400, s.c.	↓
2	stereotypy	3,6, s.c.	DDC, 500, s.c., DDC, 2 x 500, s.c. DDC, 3 x 500, s.c.	0[a]
3	locomotor activity	5, s.c.	disulfiram, 100, i.p.	↓
4	locomotor activity	4, i.p.	FLA-63, 25	0
5	locomotor activity	5, i.p.	FLA-63, 50, i.p.	↓
6	stereotypy	5, i.p.	DDC, 400, i.p.	↑[b]
7	abrupt locomotion	15, i.p.	FLA-63, 2 x 20, i.p.	↑
8	locomotor activity	1,2,4, i.p.	0.4% 4-14,624 0.05-0.1% FLA-63	0[c]
9	locomotor activity		U-14,624, 75, i.p.	0

Table 8. (cont'd) The effect of dopamine ß-hydroxylase
inhibitors on behaviors elicited by d-amphetamine administration in rats and mice

Reference	Behavior assessed	Dose of amphetamine (mg/kg)	DBH inhibitor and dose (mg/kg)	Effect on behavior
10	open field hole counts	2, i.p.	FLA-63, 25, i.p.	0 ↑
11	gnawing	5, s.c.	DDC, 400, i.p.	↑
12	stereotypy	5, i.p.	FLA-136, 15, i.p.	↑

[a] No effect was produced by any of the dosage regimens.
[b] An increase was observed at 50, 60, 100 and 120 mins. after amphetamine.
[c] No effect was observed with either drug.

References: 1, Klara-Pfeifer et al., 1966; 2, Randrup and Scheel-Kruger, 1966; 3, Maj and Przegalinski, 1967; 4, Corrodi et al., 1970; 5, Svensson, 1970; 6, Mayer and Eybl, 1971; 7, Hasselager et al., 1972; 8, Thornburg and Moore, 1973; 9, Hollister et al., 1974; 10, Ljungberg and Ungerstedt, 1976; 11, Magilnicka and Braestrup, 1976; 12, Grawboska-Anden, 1977.

3. NE Receptor Antagonists

As expected, the results regarding the effect of adrenergic antagonists on behaviors induced by apomorphine and amphetamine administration are mixed (Table 9). Thus, amphetamine-induced gnawing is potentiated by the beta-adrenoceptor antagonists dl-propranolol and MJ-1999. [Thomas and Handley, 1978] However, phenoxybenzamine potentiates, attenuates or has no effect on amphetamine and apomorphine-induced behaviors [Ljungberg and Ungerstedt, 1976; Mogilnicka and Braestrup, 1976; Thomas and Handley, 1978; Wisniowska-Szafranvec et al., 1983] In contrast, stereotyped behaviors are inhibited by the non-selective α_2 receptor antagonists yohimbine, piperoxan and rauwolscine. [Grawboska-Anden, 1977; Thomas and Handley, 1978; Zetler, 1985]

Currently, only a few studies have examined the effect of selective α_1 and α_2 receptor antagonists on amphetamine and apomorphine-induced behaviors. The systemic administration of the α_1 receptor antagonist prazosin significantly attenuates the locomotor hyperactivity induced by amphetamine. [Blanc et al., 1994; Dickinson et al., 1988; Pfil and Hornykiewicz, 1985] In addition, the increase in locomotor activity produced by the bilateral injection of d-amphetamine (4.0 nmol/side) into the nucleus accumbens was blocked by administering either prazosin or WB-4101 (0.16 pmol) into the medial prefrontal cortex. [Blanc et al., 1994] It has been reported that the selective α_2 receptor antagonist idazoxan (20 mg/kg, i.p.) produced a small but significant potentiation of d-amphetamine's action. [Dickinson et al., 1988] In contrast, prazosin enhanced the stereotyped gnawing elicited by apomorphine and amphetamine and idazoxan decreased amphetamine induced oral movements. [Dickinson et al., 1988] These results are in contrast to those reported by Luttinger and Durivage [1986] indicating that idazoxan (minimal effective dose = 1 mg/kg) effectively antagonized amphetamine-induced locomotor activity in mice. The difference between these data and those reported by Dickinson et al. [1988] could be related to the fact that different doses and species were used. In addition, if one considers that the blockade of α_2 receptors by idazoxan produces an increase in NE release, then idazoxan should potentiate amphetamine's action and prazosin should block it.

A study by Ferrari and Giuliani [1993] examined the effect of idazoxan (2 mg/kg) on various behaviors elicited by the D_2 receptor agonists BHT-920 and SND 919. Overall, idazoxan increased the stereotyped behavior produced by 2 mg/kg of BHT-920, but decreased the stereotyped induced by 1 mg/kg of SND 919.

Table 9. The effect of adrenergic antagonists on the behaviors elicited by d-amphetamine and apomorphine administration in rats and mice

Reference	Behavior measured	Challenge drug and dose (mg/kg)	Adrenergic antagonist and dose (mg/kg)	Effect on behavior
1	stereotypy	amphetamine, 3, s.c.	phentolamine, 300	0
			nethalide, 300	0
2	locomotor activity	amphetamine, 2.5, 5	phenoxybenzamine, 25	↓[a]
3	abrupt locomotion	amphetamine, 5	phenoxybenzamine, 10	↑
4	locomotor activity	apomorphine, 3, i.p.	phenoxybenzamine, 20, i.p.	0
			phentolamine 20, i.p.	↓
5	locomotor activity	amphetamine, 2	dl-propranolol (0.2–5)	↑[b]
			d-propranolol (0.5–10)	↑[c]
			oxprenolol (0.2–5)	↑[d]
			practolol (0.5–10)	↑[e]
5	stereotypy	amphetamine, 10	dl-propranolol (0.5–20)	↑[f]
			d-propranolol (1–20)	0
			oxprenolol (1–10)	0
			practolol (1–20)	0
6	gnawing	amphetamine, 5, s.c.	phenoxybenzamine, 20	↑
7	open field activity	amphetamine, 2	phenoxybenzamine, 25	0
		apomorphine, 2	phenoxybenzamine, 25	↑
8	locomotor activity	apomorphine, .25–5, s.c.	aceperone[g]	0
			phenoxybenzamine[g]	0

Table 9. (cont'd) The effect of adrenergic antagonists on the behaviors elicited by d-amphetamine and apomorphine administration in rats and mice

Reference	Behavior measured	Challenge drug and dose (mg/kg)	Adrenergic antagonist and dose (mg/kg)	Effect on behavior
9	stereotypy	amphetamine, 5, i.p.	yohimbine, 3	↓
10	gnawing	amphetamine, 12.5, i.p.	phenoxybenzamine, 20 phentolamine, 20 piperoxan, 10 yohimbine, 2 dl-propranolol, 10 MJ-1999, 10	0 ↑ ↓ ↑ ↑
11	locomotor activity	apomorphine, 0.0625, s.c.	prazosin, 0.125 yohimbine, 1.25	0 0
12	stereotypy	ambetamine, 7.5, s.c.	piperoxan, 30 WB 4101, 30 esprogyin, 30 yohimbine, 3.2[h] corynanthine, 10[h] rauwolscine, 7[h]	0 0 0 ↓ ↓ ↓
13	ambulation	amphetamine, 1	prazosin	↓
	stereotypy	amphetamine, 7	prazosin	0

Table 9. (cont'd) The effect of adrenergic antagonists on the behaviors elicited by d–amphetamine and apomorphine administration in rats and mice

Reference	Behavior measured	Challenge drug and dose (mg/kg)	Adrenergic antagonist and dose (mg/kg)	Effect on behavior
14	gnawing	amphetamine, 5, s.c.	Phenoxybenzamine 5, 10, 20, 40	→ → → →
14	gnawing	apomorphine, 5, s.c.	phenoxybenzamine 5, 10, 20, 40	→ → → →
15	locomotor activity	amphetamine, 1.8, 3.2	prazosin, 0.2; propranolol, 32	← ↑ 0
16	stereotypy	amphetamine, 6, s.c.	prazosin, 1; idazoxan, 20	→ →
17	locomotor activity	amphetamine, 0.75, i.p.	prazosin, 0.06	→
18	locomotor activity	amphetamine, 3, i.p.	Rx781094, 1[1]; rauwolscine, 1; yohimbine, 3; piperoxan, 3; tolazoline, 10	→ → → → →

Table 9. (cont'd) The effect of adrenergic antagonists on the behaviors elicited by d-amphetamine and apomorphine administration in rats and mice

Reference	Behavior measured	Challenge drug and dose (mg/kg)	Adrenergic antagonist and dose (mg/kg)	Effect on behavior
19	stereotypy	amphetamine, 10, i.p.	propranolol 5 15 50 phentolamine 5	0 →[b] →[c] 0
20	stereotypy	Amphetamine, 10, s.c.	propranolol 10 mg/kg	↓[i]

[a] Decreased with both doses of amphetamine; [b] Increased at 0.4, 1 and 3 mg/kg; [c] Increased at 1, 3, 5 and 10 mg/kg; [d] Increased at 0.5, 1 and 3 mg/kg; [e] Increased at 1 and 3 mg/kg; [f] Increased at 20 mg/kg; [g] Doses greater than 10 mg/kg had no effect; [h] These values represent ED$_{50}$, minimal effective dose. [i] ED$_{100}$ for amphetamine to increase spontaneous activity was 6.5 mg/kg; after pretreatment with propranolol, ED$_{100}$ was 25 mg/kg.

References: 1, Randrup et al., 1963; 2, Maj et al., 1972; 3, Hasselager et al., 1972; 4, Anden and Strombom, 1974; 5, Weinstock and Spriser, 1974; 6, Mogilnicka and Braestrup, 1976; 7, Ljungberg and Ungerstedt, 1976; 8, Lassen, 1976; 9, Grabwowska–Anden, 1977; 10, Thomas and Szafraniec et al., 1983; 11, Costall et al., 1981; 12, Waldmeier et al., 1982; 13, Pfil and Hornykiewicz, 1985; 14, Wiszniowska–Szafraniec; 15, Snoddy and Tessel, 1985; 16, Dickinson et al., 1988; 17, Blanc et al., 1994; 18, Luttinger and Dusivage, 1986; 19, Herman, 1967; 20, Mantegazza et al., 1968.

Table 10. The effect of noradrenergic neuronal lesions on behaviors elicited by d-amphetamine and apomorphine in rats

References	Behavior measured	Methodology	Challenge drug	Dose (mg/kg)	Effect on behavior
1	stereotypy	bilateral LC lesion with 6-OHDA	amphetamine apomorphine	0.1–10, i.p. 0.1–5, s.c.	0
2	stereotypy	bilateral, electrolytic LC lesion	amphetamine apomorphine	10, i.p. 10, s.c.	0
3	stereotypy	bilateral, electrolytic LC lesion	apomorphine	5, s.c.	0
	stereotypy	ventral NE bundle lesion			↓
4	stereotypy ambulation	50mg/kg i.p. of DSP-4	amphetamine	2, i.p. 4, i.p.	0 ↑

References: 1, Pycock, 1977; 2, Jerlicz et al., 1978; 3, Kostowski et al., 1978; 4, Ogren et al., 1983.

4. Lesions of NE Neurons.

Studies examining the effect of depletion of NE on apomorphine and amphetamine-induced behaviors has also been equivocal (Table 10). For example, the bilateral, electrolytic lesioning of the LC significantly potentiates the stereotypy elicited by 5, but not 10 mg/kg, s.c. of apomorphine but does not significantly alter d-amphetamine-induced stereotypy. [Kostowski et al., 1977, 1978] The depletion of NE using DSP-4 decreases d-amphetamine-induced locomotor activity, but did affect stereotyped behavior. [Ogren et al., 1983] Interestingly, the lesioning of the ventral NE bundle does not alter apomorphine stereotypy. [Kostowski et al., 1978] Similarly, it has been shown that the destruction of the dorsal and ventral NE projections in the rat does not alter d-amphetamine-induced locomotion. [Creese and Iversen, 1972, 1974, 1975; Roberts et al., 1975] Taken together, these results suggest that the locomotor activity induced by amphetamine is primarily mediated by DA neurons.

Although the above evidence suggests that NE may play a role in mediating or modulating hyperactive behaviors produced by amphetamine or apomorphine, the exact role remains to be elucidated for several reasons. First, a number of the compounds used such as yohimbine, piperoxan and rauwolscine lack specificity and they produce an increase in DA turnover in various brain areas, suggesting that these compounds interact with DA receptors. [Van Oene et al., 1984; Pettibone et al., 1985; Scatton et al., 1983; Waldmeier et al., 1982] Second, the administration of the various compounds could alter the pharmacokinetics of either apomorphine or amphetamine, which would confound the interpretation of the results. For example, DDC and FLA-63 have been shown to prolong the elimination of amphetamine and phenoxybenzmine lowers striatal levels of apomorphine. [Jonsson and Lewander, 1973; Wisziniowska-Szafraniec et al., 1983] Furthermore, some of the DBH inhibitors produce toxic side effects and can induce sedation and stress, thereby hampering data interpretation. [Dolphin et al., 1976; Thornburg and Moore, 1971] Indeed, it has been shown that the i.v. administration of FLA-63, which avoids the peritoneal irritation, does not alter the locomotor activation produced by d-amphetamine administration. [Corrodi et al., 1970] Third, most if not all of the agents used can have significant effects on blood pressure and produce effects such as sedation, which could alter locomotor behavior. Fourth, it is difficult to compare the studies to one another as the doses and routes of administration of the drugs were different. Fifth, it should be noted that the adrenergic agents used are not selective for the various subtypes of adrenergic receptors present in the brain.

VII. THE EFFECT OF NE ON BASAL LOCOMOTOR ACTIVITY.

There is considerable evidence that locomotor activity or ambulation in rodents is predominantly mediated by DA neurons in the brain. [For review, see Fishman et al., 1983] For example, the microinjection of DA into various mesolimbic structures produces a dose-dependent increase in motility and this effect is blocked by DA receptor antagonists. [Costall and Naylor, 1975, 1977; Jackson et al., 1975; Pijnenberg et al., 1975] However, the administration of various adrenergic antagonists does not antagonize DA-induced locomotor activity. [Costall and Naylor, 1976; Pijnenberg et al., 1975]

The microinjection of NE only produces behavioral effects in animals that have been pretreated with the monoamine oxidase inhibitor nialamide. [Costall and Naylor, 1976; Pijnenberg et al., 1975] The behavioral profile produced by NE administration consisted of motor incoordination, crawling and convulsive movements. [Costall and Naylor, 1976; Jackson et al., 1975] Interestingly, the locomotor activity produced by NE was not antagonized by adrenergic receptor antagonists. [Costall et al., 1976; Jackson et al., 1975; Pijnenberg et al., 1975] In addition, the administration of DA receptor antagonists blocked the locomotor activity produced by the intra-accumbens administration of NE. [Handley and Thomas, 1978] Thus, it appears that NE-induced locomotor activity is not the result of its interaction with NE receptors but is probably related to its interaction with DA receptors.

VIII. THE EFFECT OF CHRONIC ADMINISTRATION OF NE COMPOUNDS ON DA-MEDIATED BEHAVIORS

In the following paragraphs, we will review the data regarding the effects produced by the chronic administration of compounds that display selectivity for NE uptake sites or NE receptors on DA mediated behaviors. We will not discuss the results obtained with compounds such as imipramine and amitriptyline, as although these compounds inhibit NE uptake, they also interact with the 5-HT transporter and therefore are not selective NE uptake blockers. [Baldessarini, 1985] In addition, the major metabolites of imipramine and amitriptyline, which are desipramine and nortriptyline, respectively, block NE uptake; however, it is not known to what extent they account for the activity of the parent drugs. [Baldessarini et al., 1985]

Table 11. The effect of the chronic administration of NE uptake blockers on behaviors elicited by d-amphetamine and apomorphine in rats and mice.

Reference	Behavior measured	NE uptake blocker (mg/kg)	Length of treatment (days)	Challenge drug (mg/kg)	Effect on behavior	Species
1	locomotor activity	10, b.i.d. (+)-oxaprotaline	21	amphetamine, 1.25	↑	mice
2	stereotypy	25, desipramine	7	apomorphine, 0.25	0	rats
3	locomotor activity	5, b.i.d. desipramine	14	amphetamine 0.5 / 1 / 1.5	0 / 0 / ↑	rats
	stereotypy			2 / 5	↑ / 0	rats
4	locomotor activity	10, desipramine	10	amphetamine, 1.5	↑	rats
5	locomotor activity	10, b.i.d. desipramine	14	amphetamine 0.5 mg, NAC	↑	rats
6	locomotor activity	5, b.i.d. desipramine	21	amphetamine, 1.5	↑	rats
7	locomotor activity	10, b.i.d. desipramine, minipump	21	apomorphine 0.05	↓	rats

References: 1, Maj et al., 1983; 2, Delini-Stula et al., 1979; 3, Spyraki and Fibiger, 1981; 4, Martin-Iverson et al., 1983; 5, Maj et al., 1987; 6, Nomikos et al., 1991; 7, Allison et al., 1993.

IX. THE EFFECT OF THE CHRONIC ADMINISTRATION OF SELECTIVE NE UPTAKE BLOCKERS ON BEHAVIORS ELICITED BY D-AMPHETAMINE AND APOMORPHINE

Overall, it has been shown that the chronic administration of either oxaprotaline or desipramine produces a potentiation of amphetamine-induced locomotor activity, but not stereotypy. [Martin-Iverson et al., 1983; Maj et al., 1983; Maj et al., 1987; Nomikos et al., 1991; Spyraki and Fibiger, 1981] (Table 11) In contrast, the stereotypy elicited by apomorphine was not altered in one study whereas apomorphine-induced locomotor activity was decreased in another. [Allison et al., 1993;Delini-Stula et al., 1979] It has been postulated that the potentiation of d-amphetamine's action by desipramine may be related to the alteration of DA function in the mesolimbic area as 1) the repeated administration of desipramine does not enhance either amphetamine or apomorphine-induced stereotypy, which are believed to be mediated by the nigrostriatal DA system; 2) it has been shown that the chronic administration of desipramine produces a decrease in the number of D_1 receptors in the limbic system [Maj et al., 1987]; 3) the chronic administration of desipramine enhances the increase in the basal extracellular levels of DA in the NAc after the local or systemic administration of amphetamine [Brown et al., 1991; Nomikos et al., 1991]; 4) the chronic administration of desipramine enhances intracranial self-stimulation obtained from electrodes in the VTA, the origin of DA neurons that project to the mesolimbic system [Fibiger and Phillips, 1981]; 5) the antidepressant action of desipramine in animal models of depression is reversed by the systemic and intra-accumbens injection of D_2 antagonists. [Borsini et al., 1984; Cervo and Samanin, 1987; Delini-Stula et al., 1988; Sampson et al., 1991] However, the chronic administration of desipramine to normal rats does not alter D_2 receptor binding in either the NAc or striatum. [Lee and Tang, 1982; Martin-Iversen et al., 1983; Peroutka and Snyder, 1980; however, see Koide and Matshushita, 1981] It should be pointed out that other compounds that possess antidepressant action such as mianserin, amitriptyline, imipramine, iprindole, citalopram and buproprion also produce alterations in DA function or activity. [Chiodo and Antelman, 1980; Klimek and Nielsen, 1987; Klimek and Maj, 1989; Maj et al., 1984; Martin-Iversen et al., 1983; Serra et al., 1979] The potentiation of amphetamine's action is probably an indirect one as desipramine does not bind to DA receptors nor does it interact with the DA transporter. [Allison et al., 1993] In addition, the potentiation is unlikely related to pharmacokinetic factors as 1) acute desipramine administration does not alter amphetamine's action; 2) desipramine does not alter the accumulation of brain ^{3}H-amphetamine compared to controls; 3) it has been shown that amphetamine's action is not prolonged by the acute administration of desipramine and animals were withdrawn from desipramine for 3 days prior to amphetamine. [Spyraki and Fibiger, 1981] However, it has been previously shown that the pretreatment of rats with desipramine (2.5-5 mg/kg i.p.) markedly potentiates the levels of d-amphetamine in the brain and impairs its clearance from the brain. [Consolo et al., 1967; Sulser et al., 1966] The difference between these results and those obtained by Spyraki and Fibiger [1981] are unknown. However,the dose of amphetamine given was 7.5-15 mg/kg [Consolo et al., 1967], which is 5-10 times higher than the dose given by Spyraki and Fibiger [1981], who saw no effect of desipramine on brain levels of ^{3}H-amphetamine. In addition, the earlier studies measured whole brain levels of amphetamine or ^{3}H-amphetamine, whereas Spyraki and Fibiger [1981] measured ^{3}H-amphetamine in the cerebellum, hypothalamus, striatum, hippocampus and cortex.

It should be pointed out that imipramine and amitriptyline, which block NE and 5-HT uptake, enhance the behavioral effects produced by the microinjection of DA and amphetamine into the NAc after chronic desipramine administration. [Maj, 1986; Maj and Wedzony, 1985] In contrast, citalopram, a selective 5-HT uptake blocker, does not alter desipramine's action, suggesting that the effects of imipramine and amitriptyline are probably mediated by their blockade of NE uptake. However, there is evidence that desipramine's interaction at the NE transporter may not explain its potentiation of amphetamine's action. It has been shown that the lesioning of NE neurons with 6-OHDA or 5-HT neurons with 5,7-dihydroxytryptamine did not alter the response of animals to amphetamine after chronic desipramine treatment. [Martin-Iverson et al., 1983] Interestingly, the chronic administration of the muscarinic antagonist scopolamine increased the locomotor response to d-amphetamine. Based on this and the fact that amphetamine's action was potentiated by amitriptyline, imipramine, mianserin and iprindole (which display anticholinergic properties), but not fluoxetine, nomifensine and zimelidine, which lack anticholinergic action, it was postulated that desipramine's action may be related to its anticholinergic properties. However, Spyraki and Fibiger [1981] have shown that desipramine does not possess anticholinergic action 3 days after the last injection as assessed by measuring the effect of oxotremorine on body temperature. The explanation for the different conclusions of Martin-Iversen et al. [1983] and Spyraki and Fibiger [1981)] is unknown. It may be related to the fact that the former study administered 5, compared to 10, mg/kg of desipramine in the latter study. In addition, Martin-Iversen et al. [1983] treated animals chronically with scopolamine and assessed its effect on d-amphetamine-induced locomotion whereas Spyraki and Fibiger [1981] examined the effect of acute injections of oxotremorine on body temperature. Thus, different drug regimens, as well as muscarinic antagonists, were used.

X. THE EFFECT OF CHRONIC DESIPRAMINE ADMINISTRATION ON DA NEURONS: ELECTROPHYSIOLOGICAL STUDIES

Using the technique of extracellular single cell recording, it has been shown that the repeated administration of desipramine produces an increase in the number of spontaneously active SNc and VTA DA cells in anesthetized rats. [Chiodo and Bunney, 1983; White and Wang, 1983] The exact explanation for this phenomenon is unknown, but may be related to: 1) an alteration in the sensitivity of the D_2 autoreceptor. For example, it has been shown that the response of SNc DA cells to i.v. apomorphine, an autoreceptor agonist at low doses [Skirboll et al., 1979], is significantly attenuated after the chronic administration of the antidepressants imipramine, amitriptyline and iprindole [Chiodo and Antelman, 1980], indicating an autoreceptor subsensitivity; 2) an alteration in the pre- and postsynaptic sensitivity of NE receptors, which could subsequently alter the function of DA neurons. Thus, chronic desipramine appears to down regulate α_2 and ß adrenoreceptors, which could lead to an increase in DA function. [Caliggula and Antelman, 1977]

A study by Chiodo and Bunney [1985] has shown that the acute administration of prazosin and idazoxan produces an increase in the number of spontaneously active DA neurons in the SNc, but not in the VTA whereas the chronic administration of either compound does not alter the number of spontaneously active DA neurons in the VTA or SNc. In addition, it was shown that the chronic administration of haloperidol produced a significant decrease in the number of spontaneously active DA neurons in both areas. In contrast, the combination of haloperidol and prazosin, but not haloperidol and idazoxan, elicited a decrease and an increase in the number of spontaneously active VTA and SNc DA cells, respectively. The mechanism responsible for this effect is unknown.

XI. CATALEPSY

A number of studies have shown that catalepsy, defined as a muscular rigidity that impedes voluntary movement, may be mediated by antagonism of DA function in the nigrostriatal pathway. For example, the administration of compounds that have potent D_2 antagonists properties, such as various neuroleptics, produce catalepsy.

Table 12. The effect of noradrenergic neuronal lesioning on neuroleptic-induced catalepsy in rats

Reference	Experimental methodology	Neuroleptic administered	Dose (mg/kg)	Effect on catalepsy
1	bilateral, electrolytic LC[a] lesions[b]	haloperidol	0.1 – 2, i.p.	↑ (all doses)
2	bilateral, electrolytic LC lesion	chlorpromazine	5	↑
		haloperidol	0.5	↑
		Spiperone	1.5	↑
3	bilateral, electrolytic LC lesion	chlorpromazine	5	↑
		haloperidol	0.5	↑
	ventral NE bundle lesion	chlorpromazine	5	↓
		haloperidol	0.5	↓
4	ventral NE bundle lesion	chlorpromazine	10	↓[c]
		haloperidol	1	↓

[a] LC, locus coeruleus; [b] Lesion was performed at birth; [c] Catalepsy was completely abolished.

References: 1, Pycock et al., 1976; 2, Kostowski et al., 1977; 3, Kostowski et al., 1978; 4, Jerlicz et al., 1978.

Table 13. The effect of noradrenergic agonists and antagonists on neuroleptic-induced catalepsy in rats and mice

Reference	Neuroleptic administered	Dose (mg/kg)	Adrenergic agent (mg/kg)	Effect on catalepsy
1	haloperidol	0.1-2, i.p.	clonidine, 0.5	↑[a]
2	haloperidol	1, i.p.	phentolamine 2	0
			5	0
			10	0
			yohimbine 1	↓ (1-3 hrs)
			2	↓ (1-3 hrs)
			3	↓ (1-5 hrs)
			10	↓ (1-5 hrs)
			clonidine 0.1	↑ (2-5 hrs)
			0.5	↓ (1-3 hrs)
			phenoxybenzamine 10	0
			20	↑ (1-2 hrs)
3	haloperidol	0.2, i.p.	methoxamine, 5	↑
			clonidine, 0.01	↑
			prazosin 1	↓
			2.5	↓
			5	↑
			yohimbine 2.5	↑
			5	↑
			piperoxan, 10,20	↑,↑

References: 1, Pycock et al., 1977; 2, Al-Shabibi and Doggett, 1978; 3, Brown and Handley, 1979.

In addition, treatments which deplete striatal DA levels also induce cataleptic behavior. Experimental manipulations which alter NE neurons appear to modulate the catalepsy induced by neuroleptics (Tables 12,13). The electrolytic lesion of the LC in rats potentiates the catalepsy elicited by the neuroleptics chlorpromazine, haloperidol and spiroperidol compared to sham-lesioned controls. [Kostowski et al., 1977, 1978] In contrast, the electrolytic lesioning of the ventral NE bundle, which sends NE projections to the limbic system and hypothalamus, almost completely reduced the catalepsy produced by haloperidol and chlorpromazine. [Kostowski et al., 1978]

The results from studies examining the effect of various adrenergic compounds on haloperidol induced catalepsy have been inconsistent. For example, one study has shown that yohimbine increases, whereas another study indicates that it decreases haloperidol-induced catalepsy. [Al-Shabibi and Doggett, 1978; Brown and Handley, 1978] In contrast, clonidine decreased haloperidol-induced catalepsy in both studies. However, it is difficult to directly compare these studies as different species and doses of haloperidol were used.

It may be postulated that the potentiation of the neuroleptic-induced catalepsy by LC lesion is related to a decrease in impulse flow of DA neurons in the striatum. Thus, catalepsy would be increased as there would now be decreased levels of DA to compete with the neuroleptics for the D_2 receptors. This is corroborated by the fact that depletions of brain NE produce a 52% decrease in striatal DA levels as determined using microdialysis. Clearly, the effects of lesions of the LC, where the dorsal NE bundle and ventral NE bundle originate, were opposite. The exact explanation for this difference is unknown but could be related to the fact that the dorsal and ventral NE bundles innervate different brain areas, the main difference being the cerebral cortex, which does not receive NE innervation from the ventral NE bundle.

XII. NE/DA INTERACTIONS IN PREFRONTAL CORTEX

A number of studies conducted by Tassin and his colleagues have examined the effect of the manipulations of the NE neurons on DA function in the prefrontal cortex and ventral tegmental area and the results of these studies will be summarized below.

XIII. THE EFFECT OF NE NEURONAL LESIONING ON CORTICAL DA

Following the bilateral lesioning of NE neurons with 6-OHDA, an increase in cortical DA levels (approx 70%) was detected. [Tassin et al., 1979]   In addition, an increase in ^{3}H-DA uptake sites was observed and this was correlated with the alteration in cortical DA levels. [Tassin et al., 1979]   Subsequently, histochemical analyses indicated that there was an increase in the density of DA nerve terminals in some cortical areas following the destruction of NE neurons. Interestingly, the increase in cortical DA nerve terminals was accompanied by a 50% decrease in the DOPAC/DA ratio in the prefrontal cortices of animals devoid of NE terminals. [Tassin et al., 1979]   This decrease was not dependent upon changes in cortical DA levels as animals with almost no increase in cortical DA levels still displayed decreases of 50% in their cortical DOPAC/DA ratios. [Tassin et al., 1979]

In another study, the 6-OHDA lesions were made in the vicinity of the NE pathway connecting the LC to the VTA and corresponding to the dorsal part of the decussation of the pedunculus cerebellaris superior in the medial region of the tegmental radiation. [Herve et al., 1982]   These lesions produced a 64% decrease in NE levels in the VTA but only produced a slight decrease (19%) in NE levels in the prefrontal cortex. However, these lesions elicited a 38% decrease in the DOPAC/DA ration in the prefrontal cortex.   The NE/DA interaction in the VTA appears to be limited to mesocortical DA neurons as there was no change in DA utilization in the NAc. Overall, this suggests that NE neurons which innervate the VTA exert a specific tonic excitatory effect on mesocortical DA neurons.

XIV.   EFFECT OF CORTICAL NE INNERVATION ON REGULATION OF D_1 RECEPTORS IN THE PREFRONTAL CORTEX

There   is experimental evidence indicating that ascending NE fibers play a permissive role regarding the appearance of denervation supersensitivity of D_1 receptors in the frontal cortex. For example, the increase in the activity of DA-sensitive adenylate cyclase activity (38%) and D_1 receptor density produced in the prefrontal cortex induced by bilateral electrolytic lesions of the VTA is abolished if ascending NE neurons have been lesioned simultaneously. [Tassin et al., 1982, 1986] The lesioning of NE neurons alone did not alter adenylate cyclase activity. There appears to be a good correlation between the extent of damage to NE fibers and the reduction of cortical D_1 receptor supersensitivity as compared to that occurring in electrolytically lesioned rats. [Tassin et al., 1982a,b]   However, the bilateral administration of 6-OHDA into the VTA does not induce denervation supersensitivity of cortical D_1 receptors as this procedure destroys both ascending DA and NE fibers. [Tassin et al., 1986]

A recent study has shown that the induction of an increase in cortical adenylate cyclase activity of D_1 receptors by the irreversible alkylating agent EEDQ (0.8 mg/kg) was antagonized by the administration of 1.5 mg/kg, i.p. of the α_1 antagonist prazosin. [Trovero et al., 1992]

It has been reported that there is a hyposensitivity of ß-receptors in rats with partial lesions of NE fibers and complete lesions of cortical DA fibers. [Herve et al., 1990] This suggests that mesocortical DA neurons may regulate ß-adrenergic receptor density when the activity of ascending NE neurons is partially decreased.

XV. ROLE OF NE/DA INTERACTIONS ON THE "VENTRAL TEGMENTAL SYNDROME"

Tassin and his colleagues have performed a series of experiments examining the effect of 6-OHDA lesions of cortical NE innervation on the behavioral syndrome produced by bilateral electrolytic lesions of the VTA, which produces locomotor hyperactivity and the disappearance of spontaneous alternation. [LeMoal et al., 1976] In these experiments, animals were either lesioned bilaterally via: 1) electrocoagulations of the VTA; 2) bilateral injections of 6-OHDA into the PCS laterally; or 3) simultaneous application of both types of lesions. These experiments indicated that NE neurons play a permissive role in the expression of the behavioral deficits induced by the electrocoagulation of the VTA. Thus, while animals with only VTA lesions exhibited a 56% increase in nocturnal locomotor activity and a reduction in spontaneous alternation behavior, these changes were not observed in animals with simultaneous bilateral electrolytic lesions of the VTA and bilateral 6-OHDA lesions of the dorsal NE bundle compared to sham-operated controls. [Taghzouti et al., 1988] The lesioning of NE fibers with 6-OHDA alone did not alter locomotor activity or spontaneous alternation.

Similarly, it has been reported that the behavioral deficits produced by the lesioning of DA nerve terminals in the rat lateral septum were prevented by the lesioning of NE neurons with 6-OHDA alone [Taghzouti et al., 1991]. Therefore, the simultaneous destruction of NE neurons and DA neurons or nerve terminals markedly reduced the deficits observed in rats with electrolytic VTA or lateral septal lesions.

In another series of studies, experiments were conducted to ascertain the effect that α or ß receptor antagonists have on the behavioral deficits induced by lesioning of the VTA. [Trovero et al., 1992] As described above, the bilateral lesioning of the VTA produced an increase in behavioral hyperactivity (nocturnal locomotor activity).

The pretreatment of animals with prazosin (0.5 mg/kg, i.p.) completely abolished the locomotor hyperactivity in the lesioned rats for periods up to 24 hours after the injection. In contrast, the administration of either propranolol (5 mg/kg, i.p.) or yohimbine (1.6 mg/kg, i.p.) did not alter locomotor activity. The effect of prazosin appeared to be the result of its interaction with a subtype of the α_1 receptor as the pretreatment of animals with WB 4101, which binds to α_{1A} and α_{1B} receptors, does not modify the locomotor hyperactivity of lesioned rats. Overall, these results suggest that in vivo the stimulation of α-adrenergic receptors by NE is necessary for the expression of behavioral deficits produced by VTA lesions.

CONCLUSIONS

It is apparent that experimental manipulations which affect the activity and/or function of NE neurons in the brain can modulate the activity of DA neurons. The interpretation of some of the studies is hampered by 1) the fact that various compounds that were believed to be selective NE antagonists also possessed antidopaminergic action; 2) the administration of specific compounds that inhibited dopamine-ß-hydroxylase appear to produce stress and sedation, thus making it difficult to determine if the effects of these compounds were related to their depletion of NE or to the aforementioned properties, particularly with regards to behavioral studies; 3) the comparison of different studies was complicated by the fact that different doses and routes of drug administration were used; 4) different animal species were used; and 5) a number of the NE compounds used are not selective for specific adrenergic receptor subtypes.

Nonetheless, it appears that some general conclusions can be drawn. Overall, it appears that treatments which deplete NE levels or antagonize NE receptors can modulate the behavior in animals that have received treatments which alter DA function. For example, 1) behavioral deficits and biochemical alterations produced by the lesioning of DA neurons appear to be significantly attenuated by the destruction of NE neurons as well as by the administration of the α_1 receptor antagonist prazosin; 2) In addition, the recovery of the depletion of striatal DA levels by MPTP administration is attenuated by the bilateral lesioning of the locus coeruleus and the catalepsy induced by neuroleptic agents can be reversed by lesioning NE neurons. Thus, if the activity or function of DA neurons is compromised, it may be postulated that alterations in the activity of NE neurons or NE function may occur to compensate and that attenuation of NE function may produce an amelioration of the alterations.

However, if dopaminergic neurotransmission is intact, the effects of manipulating the NE system may have a different effect. For example, 1) the lesioning of the LC or NE nerve terminals elicits a decrease in basal extracellular DA levels in the caudate and nucleus accumbens; 2) the systemic administration of clonidine regularizes, whereas yohimbine and idazoxan deregularize the firing pattern of midbrain DA cells in normal rats; 3) the administration of NE directly into selected brain areas in normal rats does not alter locomotor activity; 4) the chronic administration of desipramine, which appears to up-regulate NE neuronal activity, generally increases the locomotion produced by d-amphetamine and alters ICSS threshold.

The interpretation of the data from some of the studies might lead one to conclude that, under normal conditions, NE may inhibit the activity in DA systems. Alternatively, the NE system may facilitate DA activity. However, a decrease in NE function may prevent or compensate the effects produced by experimental treatments that damage DA neurons or pharmacological interventions that alter DA activity. This somewhat paradoxical conclusion may be explained if, as previously proposed, NE transmission is assumed to block DA D_1 transmission in cortical areas. In this scenario, the destruction of cortical NE would then facilitate cortical DA transmission and thereby inhibit subcortical DA transmission. [Carter and Pycock, 1980; Vezina et al., 1991] In other words, intact NE function in the cortex seems necessary to obtain a normal subcortical DA transmission (Blanc et al., 1994). Thus, NE neurons would then inhibit cortical, but facilitate subcortical DA transmission.

The modulation of DA neurons by NE has clinical implications. 1) Parkinson's disease - the results from the study of Mavridis et al. [1991] suggest that the lesioning of the locus coeruleus, which produces a significant depletion of brain NE, markedly attenuates the recovery of striatal DA levels, the behavioral deficits and the number of SNc cells in animals that have received MPTP. In addition, it has been shown that the depletion of brain NE produces a decrease in the extracellular levels of DA in the nucleus accumbens and striatum. Consequently, based on this limited data, one might predict that the depletion of NE may exacerbate the symptoms of Parkinson's disease by further compromising DA function. However, as previously hypothesized, it may be predicted that the onset of Parkinson's disease may be delayed in individuals that display a significant decrease in NE function as this would serve to counteract the behavioral deficits elicited by the substantial destruction of SNc cells (and a depletion of brain DA) in this neurological disorder. [Caligula and Antelman, 1977] Clearly, due to the limited number of studies, it is difficult to ascertain which of the aforementioned hypotheses are most applicable. It may be of interest to examine the effect of various NE receptor agonists and antagonists, as well as other treatments that deplete NE on the behavioral and biochemical deficits produced by MPTP.

2) There may also be potential implications of the interaction of NE and DA with regards to schizophrenia. The dopamine theory of schizophrenia hypothesizes that the core symptoms of schizophrenia may be related to an overactivity of DA neurons or mechanisms in the mesolimbic system. [for review, see Stahl and Meltzer, 1976; Matthysse, 1973; Grace, 1991] Interestingly, there are studies indicating that the upregulation of DA function in the mesolimbic system by various experimental methods may be attenuated by treatments that attenuate NE activity or function. Thus, one might predict that the administration of NE antagonists or depletion of brain NE levels may ameliorate schizophrenic symptoms. It has been reported that a number of antipsychotic drugs are potent NE receptor antagonists. [Cohen et al., 1986] Furthermore, NE levels appear to be increased in the brains of individuals diagnosed with paranoid schizophrenia. [For review, see van Kammen and Antelman, 1984; van Kammen and Kelly, 1991] The administration of clonidine (which would attenuate NE function via activation of presynaptic α_2 receptors) to a subpopulation of schizophrenics appears to produce an improvement in symptomatology. [van Kammen et al., 1989] However, it is unlikely that treatments which only attenuate NE function alone would be effective in treating schizophrenics. Nonetheless, it is possible that a combination of such agents with standard antipsychotics may produce therapeutic response in refractory patients.

Overall, it is apparent that experimental manipulations that alter NE function can also affect the function or activity of DA neurons. However, future studies should use multiple doses of selective NE receptor subtype agonists and antagonists and various schedules of administration to aid in the interpretation of the results. Furthermore, studies that involve the administration of two drugs should examine the brain levels of each drug in order to rule out the possibility that the interpretation of the data is not hampered by pharmacokinetic factors.

REFERENCES

Al-Shabibi, M.M.H. and Doggett, N.S., On the central noradrenergic mechanism involved in haloperidol-induced catalepsy in the rat, J. Pharm. Pharmac., 30, 529, 1978.

Allison, K., Paetsch, P.R., Baker, G.B. and Greenshaw, A.J., Chronic antidepressant drug treatment attenuates motor-suppressant effects of apomorphine without changing [³H]GBR 12935 binding, Eur. J. Pharmacol., 249, 125, 1993.

Anden, N.-E., Corrodi, H., Fuxe, K., Hokfelt, B., Hokfelt. T., Rydin, C. and Svensson, T., Evidence for a central noradrenaline receptor stimulation by clonidine, Life Sci., 9, 513, 1970.

Anden, N.-E., Atack, C.V. and Svensson, T.H., Release of dopamine from central noradrenaline and dopamine nerves induced by a dopamine-ß-hydroxylase inhibitor, J. Neural Transm., 34, 93, 1973.

Anden, N.-E. and Strombom, U., Adrenergic receptor blocking agents: effects on central noradrenaline and dopamine receptors and on motor activity, Psychopharmacologia, 38, 91, 1974.

Anden, N.-E. and Grabowska, M., Pharmacological evidence for a stimulation of dopamine neurons by noradrenaline neurons in the brain, Eur. J. Pharmacol., 39, 275, 1976.

Anden, N.-E., Pauksens, K. and Svensson, K., Selective blockade of brain α_2-autoreceptors by yohimbine: effects on motor activity and on turnover of noradrenaline and dopamine, J. Neural Transm., 55, 111, 1982.

Andersson, J.L., Marcus, M., Nomikos, G.G. and Svensson, T.H., Prazosin modulates the changes in firing pattern and transmitter release induced by raclopride in the mesolimbic, but not in the nigrostriatal dopaminergic system, Naunyn-Schmied. Arch. Pharmacol., 349, 236, 1994.

Antelman, S.A. and Caggiula, A.R., Norepinephrine-dopamine interactions and behavior: a new hypothesis of stress-related interactions between norepinephrine and dopamine is proposed, Science, 195, 646, 1977.

Baldessarini, R.J., Drugs and the treatment of psychiatric disorders, in The Pharmacological Basis of Therapeutics, seventh edition, Goodman, A.G., Gilman, L.S., Rall, T.W. and Murad, F. Eds., MacMillian Publishers, New York, 1987, Chap. 19.

Berger, B., Tassin, J.-P., Blanc, G., Moyne, M.A. and Thierry, A.M., Histochemical confirmation for dopaminergic innervation of the rat cerebral cortex after destruction of the noradrenergic ascending pathways, Brain Res., 81, 332, 1974.

Blanc, G., Trovero, F., Vezina, P., Herve, D., Godeheu, A.-M., Glowinski, J. and Tassin, J.-P., Blockade of prefronto-cortical α_1-adrenergic receptors prevents locomotor hyperactivity induced by subcortical d-amphetamine injection, Eur. J. Neurosci., 6, 293-298, 1994.

Borsini, F., Nowakowska, E. and Samanin, R., Effect of repeated treatment with desipramine in the behavioral despair test in rats: antagonism by "atypical" but not classical neuroleptics or antiadrenergic drugs, Life Sci., 34, 1171, 1984.

Brown, E.E., Nomikos, G.G., Wilson, C. and Fibiger, H.C., Chronic desipramine enhances the effect of locally applied amphetamine on interstitial concentrations of dopamine in the nucleus accumbens, Eur. J. Pharmacol., 202, 125, 1991.

Brown, J. and Handley, S.L., The effects of agents acting at pre- and postsynaptic α-adrenoceptors on haloperidol catalepsy, Brit. J. Pharmacol., 42, 474P, 1979.

Bylund, D.B., Subtypes of alpha-1 and alpha-2 adrenergic receptors, FASEB J., 6, 832, 1992.

Bylund, D.B., Eikenberg, D.C., Hieble, J.P., Langer, S.Z., Lefkowitz, R.J., Minneman, K.P., Molinoff, P.B., Ruffolo, Jr., R.R. and Trendelenberg, U., IV. International Union of Pharmacology nomenclature of adrenoreceptors, Pharmacol. Rev., 46, 727, 1994.

Carboni, E., Tanda, G.L., Frau, R. and Di Chiara, G., Blockade of the noradrenaline carrier increases extracellular dopamine concentrations in the prefrontal cortex: Evidence that dopamine is taken up in vivo by noradrenergic terminals, J. Neurochem., 55, 1067, 1990.

Carter, C.J. and Pycock, C.J., Behavioral and biochemical effect of dopamine and noradrenaline depletion within the medial prefrontal cortex of the rat, Brain Res., 192, 163-176, 1980.

Cervo, L. and Samanin, R., Evidence that dopamine mechanisms in the nucleus accumbens are selectively involved in the effect of desipramine in the forced swimming test, Neuropharmacology, 10, 1469, 1987.

Chen, N.-H. and Reith, M.E.A., Effects of locally applied cocaine, lidocaine, and various uptake blockers on monoamine transmission in the ventral tegmental area of freely moving rats: A microdialysis study on monoamine interrelationships, J. Neurochem., 63, 1701, 1994

Chiodo, L.A. and Antelman, S.M., Repeated tricyclics induce a progressive dopamine autoreceptor subsensitivity independent of daily drug treatment, Nature, 287, 451, 1980.

Chiodo, L.A. and Bunney, B.S., Typical and atypical neuroleptics: differential effects of chronic administration on the activity of A9 and A10 midbrain dopaminergic neurons, J. Neurosci., 3, 1607, 1983.

Chiodo, L.A. and Bunney, B.S., Possible mechanisms by which repeated clozapine administration differentially affects the activity of two subpopulations of midbrain DA neurons, J. Neurosci., 5, 2539, 1985.

Clarke, P.B.S., Jakubovic, A. and Fibiger, H.C., Anatomical analysis of the involvement of mesolimbocortical dopamine in the locomotor stimulant actions of d-amphetamine and apomorphine, Psychopharmacology, 96, 511, 1988.

Consolo, S., Dolfini, E., Garattini, S. and Valzelli, L., Desipramine and amphetamine metabolism, J. Pharm. Pharmacol., 19, 253, 1967.

Corrodi, H., Fuxe, K., Ljungdahl, A. and Ogren, S.-O., Studies on the action of some psychoactive drugs on central noradrenaline neurons after inhibition of dopamine-ß-hydroxylase, Brain Res., 24, 451, 1970.

Costall, B. and Naylor, R., The behavioral effects of dopamine applied intracerebrally to the areas of the mesolimbic system, Eur. J. Pharmacol., 32, 87, 1975.

Costall, B. and Naylor, R., Antagonism of the hyperactivity induced by dopamine applied intracerebrally to the nucleus accumbens septi by typical neuroleptics and by clozapine, sulpiride and thioridazine, Eur. J. Pharmacol., 35, 161, 1976.

Costall, B. and Naylor, R., Differentiation of the dopamine mechanisms mediating stereotyped behavior and hyperactivity in the nucleus accumbens and caudate-putamen, J. Pharm. Pharmacol., 29, 337, 1977.

Costall, B., Marsden, C.D., Naylor, R.J. and Pycock, C.J., Stereotyped behaviour patterns and hyperactivity induced by amphetamine and apomorphine after discrete 6-hydroxydopamine lesions of extrapyramidal and mesolimbic nuclei, Brain Res., 123, 89, 1977.

Costall, B., Lim, S.K. and Naylor, R.J., Characterization of the mechanisms by which purported dopamine agonists reduce spontaneous locomotor activity of mice, Eur. J. Pharmacol., 73, 175, 1981.

Corrodi, H., Fuxe, K., Ljungdahl, A. and Ogren, S.-O., Studies on the action of some psychoactive drugs on central noradrenaline neurones after inhibition of dopamine-ß-hydroxylase, Brain Res., 24, 451, 1970.

Creese, I. and Iversen, S.D., Amphetamine response in rat after dopamine neuron destruction, Nature New Biol., 238, 247, 1972.

Creese, I. and Iversen, S.D., The role of forebrain dopamine systems in amphetamine-induced stereotyped behavior in the rat, Psychopharmacologia, 39, 345, 1974.

Creese, I. and Iversen, S.D., Pharmacological and anatomical substrates of amphetamine response in rat, Brain Res., 83, 419, 1975.

Dedek, J., Scatton, B. and Zivkovic, B., α_2 receptors are not involved in the regulation of striatal dopaminergic transmission, Brit. J. Pharmacol., 61, 361P, 1982.

Delina-Stula, A., Baumann, P. and Buch, O., Depression of exploratory activity by clonidine in rats as a model for detection of relative pre- and postsynaptic central noradrenergic receptor selectivity of α-adrenolytic drugs, Naunyn-Schmied. Arch. Pharmacol., 307, 115, 1979.

Delini-Stula, A., Raedke, E. and van Riezen, H., Enhanced functional responsiveness of the dopaminergic system - the mechanism of anti-immobility effects of antidepressants in the behavioral despair test in the rat, Neuropharmacology, 27, 943, 1988.

Dickinson, S.L., Gadie, B. and Tulloch, I.F., α_1 and α_2-adrenoceptor antagonists differentially influence locomotor and stereotyped behaviour induced by d-amphetamine and apomorphine in the rat, Psychopharmacology, 96, 521, 1988.

Dolphin, A., Sawaya, M.C.B., Jenner, P. and Marsden, C.D., Behavioural and biochemical effects of chronic reduction of cerebral noradrenaline receptor stimulation, Naunyn-Schmied. Arch. Pharmacol., 299, 167, 1977.

Duteil, J., Rambert, F.A., Pessonnier, J., Hermant, J.-F., Gombert, R. and Assous, E., Central α_1-adrenergic stimulation in relation to the behavioral stimulating effect of modafinil; studies with experimental animals, Eur. J. Pharmacol., 180, 49, 1990.

Ferrari, F. and Giuliani, D., Influence of idazoxan on the dopamine D_2 receptor agonist-induced behavioural effects in rats, Eur. J. Pharmacol., 250, 51, 1993.

Fibiger, H.C. and Philips, A.G., Increased intracranial self-stimulation in rat after long-term administration of desipramine, Science, 214, 683, 1981.

Fishman, R.H.B., Feigenbaum, J.J., Yanai, J. and Klawans, H.L., The relative importance of dopamine and norepinephrine in mediating locomotor activity, Prog. Neurobiol., 20, 55, 1983.

Foote, S.L., Bloom, F.E. and Aston-Jones, G., Nucleus locus coeruleus: new evidence of anatomical and physiological specificity, Physiol. Rev., 63, 844, 1983.

Gehlert, D.R., Gackenheimer, S.L. and Robertson, D.W., Localization of rat brain binding sites for [^{3}H]-tomoxetine, an enantiomerically pure ligand for norepinephrine uptake sites, Neurosci. Lett., 157, 203, 1993.

Geyer, M.A. and Lee, E.H.Y., Effects of clonidine, piperoxan and locus coeruleus lesion on the serotonergic and dopaminergic systems in raphe and caudate nucleus, Biochem. Pharmacol., 33, 3399, 1984.

Goldstein, M. and Nakajima, K., The effect of disulfiram on catecholamine levels in the brain, J. Pharmacol. Exp. Ther., 157, 96, 1967.

Gonon, F.G., Nonlinear relationship between impulse flow and dopamine released by rat midbrain dopaminergic neurons as studied by in vivo electrochemistry, Neuroscience, 24, 19, 1988.

Grace, A.A., Phasic versus tonic dopamine release and the modulation of dopamine system responsivity: a hypothesis for the etiology of schizophrenia, Neuroscience, 41, 1, 1991.

Grawboska-Anden, M., Modification of the amphetamine-induced stereotypy in rats following inhibition of the noradrenaline release by FLA 136, J. Pharm. Pharmacol., 29, 566, 1977.

Grenhoff, J. and Svensson, T.H., Clonidine regularizes substantia nigra dopamine cell firing, Life Sci., 42, 2003, 1988.

Grenhoff, J. and Svensson, T.H., Clonidine modulates dopamine cell firing in rat ventral tegmental area, Eur. J. Pharmacol., 165, 11, 1989.

Grenhoff, J. and Svensson, T.H., Prazosin modulates the firing pattern of dopamine neurons in rat ventral tegmental area, Eur. J. Pharmacol., 233, 79, 1992.

Grenhoff, J., Nisell, M., Ferre, S., Aston-Jones, G. and Svensson, T.H., Noradrenergic modulation of midbrain dopamine cell firing elicited by stimulation of the locus coeruleus in the rat, J. Neural Transm., 93, 11, 1993.

Hasselager, E., Rolinski, Z. and Randrup, A., Specific antagonism by dopamine inhibitors of items of amphetamine induced aggressive behavior, Psychopharmacologia, 24, 485, 1972.

Herman, Z.S., Influence of some psychotropic and adrenergic blocking agents upon amphetamine stereotyped behavior in white rats, Psychopharmacologia, 11, 136, 1967.

Herve, D., Blanc, G., Glowinski, J. and Tassin, J.-P., Reduction of dopamine utilization in the prefrontal cortex but not in the nucleus accumbens after selective destruction of noradrenergic fibers innervating the ventral tegmental area in the rat, Brain Res., 237, 510, 1982.

Herve, D., Trovero, F., Blanc, G., Vezina, P., Glowinski, J. and Tassin, J.-P., Involvement of dopamine neurons in the regulation of ß-adrenergic receptor sensitivity in rat prefrontal cortex, J. Neurochem., 54, 1864, 1990.

Hollister, A.S., Breese, G.R. and Cooper, B.R., Comparison of tyrosine hydroxylase and dopamine-ß-hydroxylase inhibition with the effects of various 6-hydroxydopamine treatments on d-amphetamine-induced motor activity, Psychopharmacologia, 36, 1, 1974.

Jackson, D.M., Anden, N.-E. and Dahlstrom, A., A functional effect of dopamine in the nucleus accumbens and in some other dopamine-rich parts of the rat brain, Psychopharmacologia, 45, 139, 1975.

Javitch, J.A., Strittmatter, S.M. and Snyder, S.H., Differential visualization of dopamine and norepinephrine uptake sites in rat brain using [3H]mazindol autoradiography, J. Neurosci., 5, 1513, 1985.

Jerlicz, M., Kostowski, W., Bidzinski, A. and Hauptmann, M., Effects of lesions in the ventral noradrenergic bundle on behavior and response to psychotropic drugs in rats, Pharmacol. Biochem. Behav., 9, 721, 1978.

Johnson, G.A., Boukma, S.J. and Kim, E.G., In vivo inhibition of dopamine-ß-hydroxylase by 1-phenyl-3-(2-thiazolyl)-2-thiourea (U-14,624), J. Pharmacol. Exp. Ther., 171, 80, 1970.

Jones, B.E. and Moore, R.Y., Ascending projections of the locus coeruleus in the rat. II. Autoradiographic studies, Brain Res., 127, 23, 1977.

Jones, B.E. and Yang, T.-Z., The efferent projections from the reticular formation and the locus coeruleus studied by anterograde and retrograde tracing in the rat, J. Comp. Neurol., 242, 56, 1985.

Jones, L.S., Gauger, L.L. and Davis, J.N., Anatomy of brain alpha-adrenergic receptors: in vitro autoradiography with [^{125}I]-HEAT, J. Comp. Neurol., 231, 190, 1985

Jonsson, J. and Lewander, T., Effects of the dopamine-ß-hydroxylase inhibitor FLA 63 on the kinetics of elimination of amphetamine in the rat, J. Pharm. Pharmac., 26, 907, 1974.

Jonsson, G., Hallman, F., Ponzio, F. and Ross, S., DSP-4: a novel compound with neurotoxic effects on noradrenergic neurons of adult and developing rats, Eur. J. Pharmacol., 72, 173, 1981.

van Kammen, D.P. and Antelman, S., Impaired noradrenergic transmission in schizophrenia, Life Sci., 34, 1403, 1984.

van Kammen, D.P. and Kelley, M., Dopamine and norepinephrine activity in schizophrenia: an integrative perspective, Schizo. Res., 4, 173, 1991.

Kehr, W., 3-methoxytyramine and normetanephrine as indicators of dopamine and noradrenaline release in mouse brain in vivo, J. Neural Transm., 50, 165, 1981.

Kelley, P.H., Seviour, P.W. and Iversen, S.D., Amphetamine and apomorphine responses in the rat following 6-OHDA lesions of the nucleus accumbens septi and corpus striatum, Brain Res., 94, 507, 1975.

Kelley, E., Jenner, P. and Marsden, C.D., Evidence that [^{3}H]-dopamine is taken up and released from nondopaminergic nerve terminals in the rat substantia nigra in vitro, J. Neurochem., 45, 163, 1985.

Kleinrok, Z., Zebrowska, I. and Wielosz, M., Some central effects of diethyldithiocarbamate administered intraventricularly in rats, Neuropharmacology, 9, 451, 1970.

Klimek, V. and Maj, J., Repeated administration of antidepressants enhances agonist affinity for mesolimbic D_2-receptors, J. Pharm. Pharmacol., 41, 555, 1988.

Klimek, V. and Nielsen, M., Chronic treatment with antidepressants decreases the number of [^{3}H]SCH 23390 binding sites in the rat striatum and limbic system, Eur. J. Pharmacol., 139, 163, 1987.

Koide, T. and Matshushita, H., An enhanced sensitivity of muscarinic cholinergic receptors associated with dopaminergic receptor subsensitivity after chronic antidepressant treatment, Life Sci., 28, 1139, 1981.

Kokkinidis, L. and Anisman, H., Amphetamine models of paranoid schizophrenia: an overview and elaboration of animal experimentation. Psychol. Bull., 88, 551, 1980.

Kostowski, W., Jerlicz, M., Bidzinski, A. and Hauptmann, M., Behavioural effects of neuroleptics, apomorphine and amphetamine after bilateral lesion of the locus coeruleus in rats, Pharmacol. Biochem. Behav., 7, 289, 1977.

Kostowski, W., Jerlicz, M., Bidzinski, A. and Hauptmann, M., Evidence for existence of two opposite noradrenergic brain systems controlling behavior, Psychopharmacology, 59, 311, 1978.

Lategan, A.J., Marien, M.R. and Colpaert, F.C., Suppression of nigrostriatal and mesolimbic dopamine release in vivo following noradrenaline depletion by DSP-4: a microdialysis study, Life Sci., 50, 995, 1992.

Lategan, A.J., Marien, M.R. and Colpaert, F.C., Effects of locus coeruleus lesions on the release of endogenous dopamine in the rat nucleus accumbens and caudate nucleus as determined by intracerebral microdialysis, Brain Res., 523, 134, 1990.

Lee, T. and Tang, S.W., Reduced presynaptic dopamine receptor density after chronic antidepressant treatment in rats, Psychiatry Res., 7, 111, 1982.

Le Moal, M., Stinus, L and Galey, D., Radio-frequency lesions of the ventral mesencephalic tegmentum: neurological and behavioural considerations, Exp. Neurol., 50, 521, 1976.

Lindvall, O., Bjorklund, A., Moore, R.Y. and Stenevi, U., Mesencephalic dopamine neurons projecting to neocortex, Brain Res., 81, 325, 1974.

Ljungberg, T. and Ungerstedt, U., Automatic registration of behavior related to dopamine and noradrenaline transmission, Eur. J. Pharmacol., 36, 181, 1976.

Luttinger, D. and Durivage, M.E., Alpha$_2$-adrenergic antagonists effect on amphetamine-induced behavior, Pharmacol. Biochem. Behav., 25, 155, 1986.

MacG. Donaldson, I., Dolphin, A., Jenner, P., Marsden, C.D. and Pycock, C., The roles of noradrenaline and dopamine in contraversive circling behaviour seen after unilateral electrolytic lesions of the locus coeruleus, Eur. J. Pharmacol., 39, 179, 1976.

McKenzie, G.M. and Soroko, F.E., Inhibition of the anticonvulsant activity of L-dopa by FLA-63, a dopamine-ß-hydroxylase inhibitor, J. Pharm. Pharmacol., 25, 76, 1973.

Maj, J., Sowinska, H., Kapturkiewicz, Z. and Sarnek, J., The effect of L-DOPA and (+)-amphetamine on the locomotor activity after pimozide and phenoxybenzamine, J. Pharm. Pharmacol., 24, 412, 1972.

Maj, J., Sowinska, H., Baran, L. and Kapturiewcz, Z., The effect of clonidine on locomotor activity in mice, Life Sci., 11, 483, 1972.

Maj, J. and Przegalinski, E., Disulfiram and some effects of amphetamine in mice and rats, J. Pharm. Pharmacol., 19, 341, 1967.

Maj, J., Przegalinski, E. and Wielosz, M., Disulfiram and the drug-induced effects on motility, J. Pharm. Pharmacol., 20, 247, 1968.

Maj, J., Skuza, G. and Sowinska, H., Repeated treatment with antidepressant drugs potentiates the locomotor response to (+)-amphetamine, J. Pharm. Pharmacol., 36, 127, 1984.

Maj, J. and Wedzony, K., Repeated treatment with imipramine or amitriptyline increase the locomotor response of rats to (+)-amphetamine given into the nucleus accumbens, J. Pharm. Pharmacol., 37, 362, 1985.

Maj, J., Wedzony, K. and Klimek, V., Desipramine given repeatedly enhances behavioural effects of dopamine and d-amphetamine injected into the nucleus accumbens, Eur. J. Pharmacol., 140, 179, 1987.

Maj, J., Repeated treatment with antidepressant drugs: responses mediated by brain DA receptors, in New Results in Depression Research, Hippius, H., Klerman, G.L. and Matussek, N., Eds., Springer-Verlag, Berlin, 1986, pg. 90.

Makanjuola, R.O.A., Dow, R.C. and Ashcroft, G.W., Behavioural responses to stereotactically controlled injections of monoamine neurotransmitters into the accumbens and caudate-putamen nuclei, Psychopharmacology, 71, 227, 1980.

Mantegazza, P., Kabir, M., and Riva, M., Effects of propranolol on some activities of amphetamine, Eur. J. Pharmacol., 4, 25, 1968.

Martin-Iverson, M.T., LeClere, J.-F. and Fibiger, H.C., Cholinergic-dopaminergic interactions and the mechanisms of action of antidepressants, Eur. J. Pharmacol., 94, 193, 1983.

Matthysse, S., Antipsychotic drug actions: a clue to the neuropathology of schizophrenia? Fed. Proc., 32, 200, 1973.

Mavridis, M., Degryse, A.-D., Lategan, A.J., Marien, M.R. and Colpaert, F.C., Effects of locus coeruleus lesions on parkinsonian signs, striatal dopamine and substantia nigra cell loss after 1-methyl-4-phenyl-1,2,3,6-tetrahydropyridine in monkeys: a possible role for the locus coeruleus in the progression of parkinson's disease, Neuroscience, 41, 507, 1991.

Mayer, O. and Eybl, V., The effect of diethyldithiocarbamate on amphetamine-induced behavior in rats, J. Pharm. Pharmac., 23, 894, 1971.

Meltzer, H.Y. and Stahl, S.M., The dopamine hypothesis of schizophrenia: A review, Psychopharmacol. Bull., 2, 19, 1976.

Mogilnicka, E. and Braestrup, C., Noradrenergic influence on the stereotyped behavior induced by amphetamine, phenethylamine and apomorphine, J. Pharm. Pharmacol., 28, 253, 1976.

Molander, L. and Randrup, A., Effects of thymoleptics on behavior associated with changes in brain dopamine. II. Modification and potentiation of apomorphine-induced stimulation of mice. Psychopharmacology, 49, 139, 1976.

Mueller, K. and Nyhan, W.L., Modulation of the behavioral effects of amphetamine in rats by clonidine, Eur. J. Pharmacol., 83, 339, 1982.

Nomikos, G.G., Damsma, G., Wenkstern, D. and Fibiger, H.C., Chronic desipramine enhances amphetamine-induced increases in interstitial concentrations of dopamine in the nucleus accumbens, Eur. J. Pharmacol., 195, 63, 1991.

Nurse, B., Russell, V.A. and Taljaard, J.J.F., Effect of chronic desipramine treatment on adrenoceptor modulation of [^{3}H]Dopamine release from rat nucleus accumbens slices, Brain Res., 334, 235, 1985

van Oene, J.C., Houwig, H.A. and Horn, A.S., Evidence that the purported dopaminergic agonist (3,4-dihydroxyphenylimino)-2-imidazolidine (DPI) may reduce rat striatal dopamine turnover by an α_2-adrenergic mechanism, Eur. J. Pharmacol., 81, 75, 1982.

van Oene, J.C., de Vries, J.B. and Horn, A.S., The effectiveness of yohimbine in blocking rat central dopamine autoreceptors in vivo, Naunyn-Schmied. Arch. Pharmacol., 327, 304, 1984.

Ogren, S.O., Archer, T. and Johansson, C., Evidence for a selective brain noradrenergic involvement in the locomotor stimulant effects of amphetamine in the rat, Neurosci. Lett., 43, 327, 1983.

Papeschi, R. and Theiss, P., The effect of yohimbine on the turnover of brain catecholamines and serotonin, Eur. J. Pharmacol., 33, 12, 1975.

Peroutka, S.J. and Snyder, S.H., Long-term antidepressant treatment decreases spiroperidol-labeled serotonin receptor binding, Science, 88, 88, 1980.

Persson, T., Drug-induced changes in ^{3}H-catecholamine accumulation after ^{3}H-tyrosine, Acta Pharmacol. Toxicol., 28, 378, 1970.

Pettibone, D.J., Pfleuger, A.B. and Totaro, J.A., Comparison of the effects of recently developed α_2-adrenergic antagonists with yohimbine and rauwolscine on monoamine synthesis in rat brain, Biochem. Pharmacol., 34, 1093.

Pfeifer, A.K., Galambos, E. and Gyorgy, L., Some central nervous system properties of diethyldithiocarbamate, J. Pharm. Pharmac., 18, 254, 1966.

Phil, C.F. and Hornykiewicz, O., Alpha-noradrenergic involvement in locomotor activity, Naunyn-Schmied. Arch. Pharmacol., 33, R71, 1985.

Phillipson, O.T., Afferent projections to the ventral tegmental area of Tsai and interfascicular nucleus: a horseradish peroxidase study in the rat, J. Comp. Neurol., 187, 117, 1979.

Pijnenberg, A.J.J., Honig, W.M.M. and Van Rossum, J.M., Effects of antagonists upon locomotor stimulation induced by injection of dopamine and noradrenaline into the nucleus accumbens of nialamide-pretreated rats, Psychopharmacologia, 41, 175, 1975.

Pycock, C.J., MacG. Donaldson, I. and Marsden, C.D., Circling behavior produced by unilateral lesions in the region of the locus coeruleus in rats, Brain Res., 97, 317, 1975.

Pycock, C.J., Jenner, P.G. and Marsden, C.D., The interaction of clonidine with dopamine-dependent behaviour in rodents, Naunyn-Schmied. Arch. Pharmacol., 297, 133, 1977.

Rabey, J.M., Passeltiner, P., Bystritsky, Engel, K. and Goldstein. M., The regulation of striatal DOPA synthesis by α_2 adrenoreceptors, Brain Res., 230, 422, 1981.

Randrup, A., Munkvad, I. and Udsen, P., Adrenergic mechanisms and amphetamine induced abnormal behaviors, Acta Pharmacol. et Toxicol., 20, 145, 1963.

Randrup, A., Scheel-Kruger, J., Diethyldithiocarbamate and amphetamine stereotyped behavior, J. Pharm., 18, 752, 1966.

Reches, A. and Meiner, Z., The locus coeruleus and dopaminergic function in rat brain: implications to parkinsonism, Brain Res. Bull., 28, 663, 1992.

Reisine, T.D., Chesselet, M.F., Lubetzki, C., Cheramy, A. and Glowinski, J., A role for striatal beta-adrenergic receptors in the regulation of dopamine release, Brain Res., 241, 123, 1982.

Roberts, D.C.S., Zis, A.P. and Fibiger, H.C., Ascending catecholamine pathways and amphetamine-induced locomotor activity: importance of dopamine and apparent non-involvement of norepinephrine, Brain Res., 93, 441, 1975.

Rolinski, Z. and Scheel-Kruger, J., The effect of dopamine and noradrenaline antagonists on amphetamine-induced locomotor activity in mice and rats, Acta Pharmacol. et Toxicol., 33, 385, 1973.

Russell, V.A., Lamm, M.C.L. and Taljaard, J.J.F., Lack of interaction between α_2-adrenoceptors and dopamine D_2-receptors in mediating their inhibitory effects on [^{3}H]dopamine release from rat nucleus accumbens slices, Neurochem. Res., 18, 285, 1993.

Sampson, D., Muscat, R. and Willner, P., Reversal of antidepressant action by dopamine antagonists in an animal model of depression, Psychopharmacology, 104, 491, 1991.

Scatton, B., Zivkovic, B. and Dedek, J., Antidopaminergic properties of yohimbine, J. Pharmacol., Exp. Ther., 215, 494, 1980.

Scatton, S., Dedek, J. and Zivkovic, B., Lack of involvement of α_2 adrenoceptors in the regulation of striatal dopaminergic transmission, Eur. J. Pharmacol., 86, 427, 1983.

Scheinin, M., Lomasney, J.W., Hayden-Hixson, D.M., Schambra, U.B., Caron, M.G., Lefkowitz, R.J. and Fremeau, Jr., R.T., Distribution of α_2-adrenergic receptor subtype gene expression in rat brain, Mol. Brain Res., 21, 133, 1994.

Serra, G., Argiolas, A., Klimek, V., Fadda, F. and Gessa, G.L., Chronic treatment with antidepressants prevents the inhibitory effect of small doses of antidepressants of apomorphine on dopamine synthesis and motor activity, Life Sci., 415, 1979,

Spyraki, C. and Fibiger, H.C., Behavioural evidence for supersensitivity of postsynaptic dopamine receptors in the mesolimbic system after chronic administration of desipramine, Eur. J. Pharmacol., 74, 195, 1981.

Strombom, U., Catecholamine receptor agonists. Effects on motor activity and rate of tyrosine hydroxylation in mouse brain, Naunyn-Schmied. Arch. Pharmacol., 292, 167, 1976.

Sulser, F., Owens, M.L. and Dingell, J.V., On the mechanism of amphetamine potentiation by desipramine (DMI), Life Sci., 5: 2005, 1966.

Svensson, T.H., The effect of inhibition of catecholamine synthesis on dexamphetamine induced central stimulation, Eur. J. Pharmacol., 12, 161, 1970.

Svensson, T.H. and Waldeck, B., On the significance of central noradrenaline for motor activity: experiments with new dopamine-ß-hydroxylase inhibitor, Eur. J. Pharmacol., 7, 278-282, 1969.

Svensson, T. and Waldeck, B., On the role of catecholamines in motor activity: experiments with inhibitors of synthesis and monoamine oxidase, Psychopharmacology, 18, 357, 1970.

Taghzouti, K., Simon, H., Herve, D., Blanc, G., Studler, J.M., Glowinski, J., Le Moal, M. and Tassin, J.-P., Behavioural deficits induced by an electrolytic lesion of the rat ventral mesencephalic tegmentum are corrected by a superimposed lesion of the dorsal noradrenergic system, Brain Res., 440, 172, 1988.

Taghzouti, K., Le Moal, M. and Simon, H., Suppression of noradrenergic innervation compensates for behavioral deficits induced by lesion of dopaminergic terminals in the lateral septum, Brain Res., 552, 124, 1991.

Tassin, J.-P., Lavielle, S., Herve, D., Blanc, G., Thierry, A.M., Alvarez, C., Berger, B. and Glowinski, J., Collateral sprouting and reduced activity of the rat mesocortical dopaminergic neurons after selective destruction of the ascending noradrenergic bundles, Neuroscience, 4, 1569, 1979.

Tassin, J.-P., Simon, H., Herve, D., Blanc, G., Le Moal, M., Glowinski, J. and Bockaert, J., Non-dopaminergic fibers may regulate dopamine-sensitive adenylate cyclase in the prefrontal cortex and nucleus accumbens, Nature, 295, 696, 1982a.

Tassin, J.P., Simon, H., Glowinski, J. and Bockaert, J., Modulations of the sensitivity of DA receptors in the prefrontal cortex and the nucleus accumbens: relationship with locomotor activity, in Brain peptides and hormones, Collu, R., Ducharme, J.R., Barbeau, A. and Tobis, B., Eds., Raven Press, New York, 1982b, pp. 17-30.

Tassin, J.P., Studler, J.M., Herve, D., Blanc, G. and Glowinski, J., Contribution of noradrenergic neurons to the regulation of dopaminergic (D1) receptor denervation supersensitivity in rat prefrontal cortex, J. Neurochem., 46, 243, 1986,

Tassin, J.-P., NE/DA interactions in prefrontal cortex and their possible roles as neuromodulators in schizophrenia, J. Neural Transm., 36, 135, 1992.

Thierry, A.M., Blanc, G., Sobel, A., Stinus, L. and Glowinski, J., Dopaminergic terminals in the rat cortex, Science, 182, 499, 1973.

Thomas, K.V. and Handley, S.L., Modulation of dexamphetamine-induced compulsive gnawing - including the possible involvement of presynaptic alpha-adrenoceptors, Psychopharmacology, 56, 61, 1978.

Thornburg, J.E. and Moore, K.E., Stress-related effects of various inhibitors of catecholamine synthesis in the mouse, Arch Int. Pharmacodyn., 194, 158, 1971.

Thornburg, J.E. and Moore, K.E., The relative importance of dopaminergic and noradrenergic neuronal systems for the stimulation of locomotor activity induced by amphetamine and other drugs, Neuropharmacology, 12, 853, 1973.

Trovero, F., Blanc, G., Herve, D., Vezina, P., Glowinski, J. and Tassin, J.-P., Contribution of an α_1-adrenergic receptor subtype to the expression of the "ventral tegmental area syndrome", Neuroscience, 47, 69, 1992.

Ueda, H., Goshima, Y. and Misu, Y., Presynaptic mediation by α_2-, β_1- and β_2-adrenoceptors of endogenous noradrenaline and dopamine release from slices of rat hypothalamus, Life Sci., 4, 371, 1983.

Vezina, P., Blanc, G., Glowinski, J. and Tassin, J.-P., Opposed behavioral outputs of increased dopamine transmission in prefrontocortical and subcortical areas: a role for cortical D1 receptors, Eur. J. Neurosci., 1001, 1991.

Von Voigtlander, P.F. and Moore, K.E., Behavioral and brain catecholamine depleting actions of U-14,624, an inhibitor of dopamine ß-hydroxylase, Fed. Proc., 53, 817, 1973.

Waldmeier, P.C., Ortmann, R. and Bischoff, S., Modulation of dopaminergic transmission by alpha-noradrenergic agonists and antagonists: evidence for antidopaminergic properties of some alpha antagonists, Expeientia, 38, 1168, 1982.

Walter, D.S., Flockhart, I.R., Haynes, M.J., Howlett, D.R., Lane, L.C., Burton, R., Johnson, J. and Dettmar, P.W., Effects of idazoxan on catecholamine systems in rat brain, Biochem. Pharmacol., 33, 2553, 1984.

Weinstock, M. and Speiser, Z., Modification by propranolol and related compounds of motor activity and stereotype behaviour induced in the rat by amphetamine, Eur. J. Pharmacol., 25, 29, 1974.

White, F.J. and Wang, R.Y., Differential effects of classical and atypical antipsychotic drugs on A9 and A10 dopamine neurons, Science, 221, 1054, 1983.

Wiszniowska-Szafraniec, G., Danek, L., Reichenberg, K. and Vetulani, J., Facilitation by α-adrenolytics of apomorphine gnawing behavior: depression of threshold apomorphine concentration in the striatum of the rat, Pharmacol. Biochem. Behav., 19, 19, 1983.

Worth, W.S., Collins, J., Kett, D. and Austin, J.H., Serial changes in norepinephrine and dopamine in rat brain after locus coeruleus lesions, Brain Res., 106, 198-203, 1976.

Zetler, G., Clonidine sensitizes mice for apomorphine-induced stereotypic gnawing: antagonism by neuroleptic and cholecystokinin-like peptides, Eur. J. Pharmacol., 111, 309, 1985.

CHAPTER 2

DOPAMINE-OPIOID INTERACTIONS IN THE BASAL FOREBRAIN

Lynn Churchill and Peter W. Kalivas

ABBREVIATIONS

6-OHDA--6-hydroxydopamine
AMP--adenosine monophosphate
AP--activated protein
CREB--cAMP response element binding
D#--dopamine receptor subtype
DA--dopamine
DAMGO--Tyr-D-Ala-Gly-NmePhe-Gly-OH
DNA--deoxyribonucleic acid
DPDPE--[D-penicillamine^{2,5}]-enkephalin
EM--electron microscope
G--guanine nucleotide binding
GABA--γ-aminobutyric acid
GAD--glutamic acid decarboxylase
IR--immunoreactive
μ,δ,κ--mu,delta or kappa opioid
mRNA--messenger ribonucleic acid
PKA--cAMP dependent protein kinase
PKC--protein kinase C
TH--tyrosine hydroxylase
VTA--ventral tegmental area

Additional figure abbreviations
AC--adenylyl cyclase
ENK--enkephalin
MD--mediodorsal thalamus
mPFC--medial prefrontal cortex
NAcc--nucleus accumbens
VP--ventral pallidum

INTRODUCTION

An interaction between dopamine and opioid transmission in the brain is perhaps most clearly evident by the fact that both indirect dopamine agonists, such as cocaine and amphetamine, and opioid agonists, such as heroin and morphine, alter the emotional and motivational state of humans resulting in abuse of these drugs. A relatively lucid literature has emerged indicating overlap in the anatomical and neurochemical substrates which support reinforcement-based behavioral responses to both classes of drugs. Furthermore, in species where opioid agonists elicit motor stimulation there is overlap in the neural substrates mediating behavioral activation elicited by systemic administration of opioids and indirect dopamine agonists. Based upon these pharmacological data an interaction between endogenous opioids and dopamine has been proposed as an important component in the physiological and pathological regulation of mood and motivation. This hypothesis will be amplified herein by critically evaluating the anatomical, biochemical and behavioral evidence supporting an interaction between dopamine and enkephalin transmission.

I. ANATOMICAL EVIDENCE FOR A
DOPAMINE-OPIOID INTERACTION

Dopamine neurons in the ventral mesencephalon project to numerous cortical, limbic and motor nuclei. [Fallon and Moore, 1978; Swanson, 1982] The dopamine system which has been most clearly identified as a neural substrate where dopamine and opioids interact to regulate the motivational state of mammals is the projection from the ventral tegmental area (VTA) to the nucleus accumbens. [Koob et al., 1993; Wise and Rompre, 1989] In part by virtue of its role in regulating motivational behaviors, the mesoaccumbens dopamine system has been hypothesized to "gate" the transfer of information between classic limbic and motor nuclei. [Mogenson, 1987; LeMoal and Simon, 1991] Thus, the status of mesoaccumbens dopamine transmission is proposed to be a prime determinant in regulating the initiation of a behavioral response to environmental stimuli. [LeMoal and Simon, 1991; Mogenson et al., 1993] In performing this function, the mesoaccumbens dopamine projection is embedded in a topographically organized circuit that consists of numerous positive and negative feedback relationships. These interconnections permit simultaneous access of the mesoaccumbens system to many brain nuclei previously identified as playing a decision-making role in regulating motivational behavioral responses, including the amygdala, prefrontal cortex, ventral pallidum and mediodorsal thalamus. [Swerdlow and Koob, 1987; Zahm and Brog, 1992; Kalivas et al., 1993a] This circuit has previously been described as the "motive circuit" and Figure 1 illustrates the location of dopamine and endogenous opioid projections within the motive circuit.

Dopamine projects from the VTA to all of the individual nuclei in the motive circuit. [Swanson, 1982] While opioid peptides and/ or receptors are also present in greater or lesser amounts in all nuclei, the projections that have been characterized to date are those originating in the nucleus accumbens and projecting to the ventral pallidum and VTA. [Zahm et al., 1985; Kalivas et al., 1993b] While further data are necessary to fully evaluate the anatomical organization of opioid peptides within the motive circuit, there is a substrate permitting endogenous opioid modulation of dopamine transmission. Moreover, it is possible that the modulation may be at the level of both dopamine perikarya in the VTA and axon terminal fields, and may affect both presynaptic dopamine release and the postsynaptic actions of released dopamine.

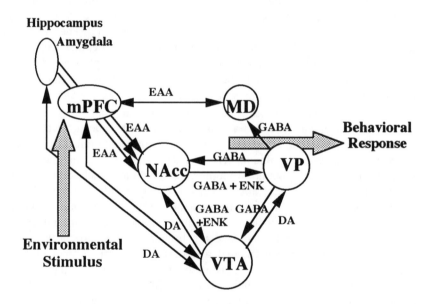

Fig. 1. A schematic drawing of the "motive" circuit which contains brain nuclei interconnected with the mesoaccumbens projection that are involved in gating the translation of motivationally-relevant environmental stimuli into adaptive behavioral responses. The hippocampus, amygdala and medial prefrontal cortex (mPFC) receive the environmental stimuli and project to the nucleus accumbens (NAcc) and VTA using mainly excitatory amino acids (EAA). The NAcc interconnects with the mesolimbic locomotor circuit which includes the ventral pallidum (VP) and ventral tegmental area (VTA) using GABA and opioids. The projection out from the VP through the mediodorsal thalamus (MD) which is partially GABAergic feeds back onto the mPFC using the excitatory amino acids (see, [Zahm and Brog, 1992; Kalivas et al., 1993a] for a more detailed description of this circuit).

A. CELLULAR ORGANIZATION WITHIN THE MESOLIMBIC CIRCUITRY

An anatomical catalogue of the various sites where DA and opioids may interact reveals a large number of possibilities.   The presence of three endogenous opioid systems, proopiomelanocortin, enkephalin and dynorphin, as well as three types of receptors--μ, δ, and κ receptors, with each receptor having at least 2 subtypes [Mansour et al., 1987; Mansour and Watson, 1993] and 5 subtypes for the dopamine receptor [Surmeier et al., 1993], implies that there exists numerous possibilities for interaction.   Another complexity in studying this interaction is that each of the endogenous opioid precursors can be metabolized into numerous fragments and none of the fragments is absolutely selective for a single opioid receptor subtype. [Kosterlitz, 1991]   All three opioid receptors and peptides are prominent in the nucleus accumbens.   The three opioid receptor subtypes have moderate to dense binding which is associated with either enkephalin or dynorphin cells and fibers as well as preopiomelanocortin fibers. [Mansour and Watson, 1993]   The μ opioid receptors form patches that appear to line up with dynorphin peptide distribution rather than enkephalin.   In contrast, the δ and κ opioid receptors are diffusely distributed with a mediolateral topography.   The δ receptors are more dense laterally, while the κ receptors present the inverse topography.   Although both enkephalin and dynorphin fibers densely innervate the ventral pallidum, the concentration of opioid receptors are light relative to the nucleus accumbens. Enkephalin fibers have been demonstrated to project from the nucleus accumbens to the ventral pallidum [Haber and Nauta, 1983; Zaborszky et al., 1985; Zahm et al., 1985] but not from the ventral pallidum back to the nucleus accumbens. [Churchill and Kalivas, 1994]   Using in situ hybridization to estimate the density of preproenkephalin mRNA-containing neurons, approximately 50% of the medium spiny neurons in the nucleus accumbens are enkephalinergic.   In the VTA and medial substantia nigra compacta where the dopaminergic neurons are concentrated, a scattering of preopiomelanocortin and enkephalin fibers are evident along with moderate levels of μ opioid receptor binding and low amounts of δ and κ receptors. [Mansour and Watson, 1993] Retrograde labeling studies reveal that neurons in the medial nucleus accumbens containing preproenkephalin mRNA project to the VTA. [Kalivas et al., 1993b]

Recent studies on molecular cloning of the dopamine receptors have revealed five subtypes as well as alternate splicing for some of the D_2 subtype, but the pharmacological distinction for each of these subtypes has been difficult to demonstrate (see review, [Sibley and Monsma, 1992]).   Basically, two families of receptors still preside with the D_{1A} and D_{1B} (D_5) subtypes in the D_1 family and the D_{2-4} subtypes in the D_2 family. [Seeman and Van Tol, 1993] The anatomical localization of the dopamine receptor subtypes has demonstrated a high concentration in the terminal fields of the dopaminergic projection for the D_{1A}, D_2 and D_3 subtypes.   The D_{1A} and D_2 receptors are present in high concentrations throughout the nucleus accumbens and caudate-putamen [Mansour et al., 1990; Joyce et al., 1991; Le Moine et al., 1991], whereas the D_3

receptor mRNA and binding sites are in high concentrations in the islands of Calleja and in lower concentrations in the nucleus accumbens. [Levesque et al., 1992] The D_4 receptor subtype appears to have a cortical, limbic and hypothalamic distribution [Van Tol et al., 1991], whereas the D_{1B} (D_5) receptor subtype is localized specifically in the hippocampus and thalamus. [Meador-Woodruff et al., 1992] In the ventral mesencephalon, the presence of both D_{1A} and D_2 receptor subtypes has been demonstrated. The D_2 receptor mRNA is localized in the dopaminergic cell bodies within both the substantia nigra and VTA. [Mansour et al., 1990] In contrast, the D_{1A} receptors are localized mainly on afferent terminals, with no D_{1A} mRNA present in the VTA. [Mansour et al., 1991]

Dopamine receptor subtypes are localized on opioidergic neurons within the mesolimbic circuit suggesting that these neurons are being regulated by dopamine input. Using in situ hybridization cytochemistry, projection neurons in the nucleus accumbens and striatum containing preproenkephalin A mRNA express mRNA primarily for D_2 dopamine receptors [Gerfen et al., 1990; Le Moine et al., 1991] and in the striatum those containing prodynorphin mRNA contain mRNA primarily for D_{1A} dopamine receptors. Furthermore, enkephalinergic neurons project more selectively to the globus pallidus while the dynorphin cells contribute primarily to the striatonigral projection. However, these conclusions based upon in situ hybridization studies only demonstrate the somatic abundance of the mRNAs, which may not be related to the level of transcribed protein. [Surmeier et al., 1993] The more sensitive technique of polymerase chain reaction, as well as electrophysiological studies, reveals that most neostriatal projection neurons contain both D_1 and D_2 dopamine receptor mRNAs. The D_2 receptor protein is found in 60% of the striatonigral neurons, demonstrating that even though the D_2 receptor mRNA is in relatively lower abundance, the transcription of D_2 receptors is still evident. [Ariano et al., 1992] However, the fact that a selective lesion of striatopallidal neurons decreases D_2 receptor binding but not D_1 receptor binding in the globus pallidus suggests that the striatopallidal neurons containing preproenkephalin mRNA do have a higher concentration of D_2 than D_1 receptors. [Harrison et al., 1992] Furthermore, pharmacological manipulations of the striatum with specific D_1 or D_2 antagonists or agonists elicit selective changes in opioid peptides in either the striatonigral or striatopallidal neurons, respectively, suggesting that the receptor types may be regulating distinct projection pathways. [Robertson et al., 1992a]

These data argue that the opioid projections from the striatum are functionally organized according to terminal field and dopamine receptor subtypes which influence their activity. Whether these distinctions apply for the opioid projections from the nucleus accumbens to pallidal and ventral mesencephalic terminal fields remains to be verified. Based upon the existing data it appears that the distinctions observed in striatal projections may not be as pronounced in the nucleus accumbens. For example, while less dense than the projection to the ventral pallidum, an enkephalinergic projection from the nucleus accumbens to the VTA has been identified. [Kalivas et al., 1993b]

Furthermore, the alterations in opioid transmission produced by dopamine antagonists or agonists in striatal projections are generally less marked in the accumbens and its pallidal and mesencephalic terminal fields. [Reiner and Andersen, 1990]

B. SYNAPTIC ORGANIZATION WITHIN THE MESOLIMBIC CIRCUITRY

Electron microscopic (EM) analysis of enkephalin immunoreactive cell bodies and terminals within the striatum demonstrate that dopaminergic terminals synapse on enkephalinergic cells adjacent to enkephalinergic terminals. [Pickel et al., 1992] These enkephalinergic terminals appear to be derived from collaterals of the same or neighboring spiny neurons. [Izzo et al., 1987] Besides the monosynaptic input of dopaminergic afferents onto enkephalinergic neurons and the apposition of dopaminergic and enkephalinergic afferents without intervening astrocytes, both afferents interact by converging onto common targets. Although EM analysis of enkephalin immunoreactive neurons has not been performed in the nucleus accumbens, the fact that enkephalin- and glutamic acid decarboxylase (GAD; synthetic enzyme for GABA)-immunoreactive neurons colocalize [Zahm et al., 1985; Besson et al., 1990] and that catecholaminergic afferents synapse on medium spiny projection neurons with GABA immunoreactivity in the medial accumbens [Pickel et al., 1988] suggests that the dopaminergic afferents synapse onto GABAergic projection neurons that also contain enkephalin. This possibility is further supported by the observation that all of the neurons containing preproenkephalin A mRNA also contain the mRNA for the D_2 dopamine receptor in both the nucleus accumbens and striatum, which implies that dopamine influences the enkephalinergic neurons through this receptor. [Le Moine et al., 1991]

In the ventral tegmental area, enkephalin-IR terminals synapse directly on tyrosine hydroxylase (TH)-IR dendrites as well as nondopaminergic dendrites with mainly symmetric contacts. [Sesack and Pickel, 1992; Liang et al., 1993] Presynaptic regulation by enkephalin is also supported by the presence of direct axonal appositions without astrocytes between enkephalin-IR terminals and other terminals, as well as the localization of the enkephalin-immunoreactivity in large dense core vesicles that have been shown to involve non-synaptic release in other systems (see review, [Thureson-Klein and Klein, 1990]). Also the enkephalin in these terminals may be colocalized with other transmitters, such as GABA, as evidenced by the similarity in the type of synaptic contacts [Bayer and Pickel, 1991] and preliminary evidence for colocalization. [Sesack and Pickel, 1992]

C. ANATOMICAL MODIFICATIONS IN ENKEPHALINERGIC TERMINALS OR NEURONS AFTER DOPAMINE DEPLETION

The strongest evidence that the opioids and dopamine systems are interacting comes from the experiments where dopaminergic innervation is depleted by the neurotoxin, 6-hydroxydopamine, or dopamine receptors are blocked by dopamine antagonists. Unilateral lesions of the dopaminergic neurons in the ventral mesencephalon increase the tissue content of enkephalin and mRNA for preproenkephalin in the striatum [Voorn et al., 1987; Vernier et al., 1988] as well as decreasing the peptide and mRNA levels for dynorphin [Gerfen et al., 1991] but does not appear to increase mRNA levels in the nucleus accumbens. [Cenci et al., 1993] These changes can be normalized by intrastriatal fetal dopaminergic transplants. Also, the decreases in prodynorphin mRNA which were observed in the striatum after the unilateral 6-hydroxydopamine (6-OHDA) lesions of the dopaminergic pathway were not significant in the nucleus accumbens. EM analysis suggests that the enkephalinergic terminals in the striatum enlarge in size and length of the synaptic specialization after dopamine depletion. [Ingham et al., 1991] These morphological alterations might enhance the synaptic efficacy of enkephalin transmission. Repeated treatment with dopamine blockers also increases the level of enkephalin peptide and the mRNA content in both the striatum and nucleus accumbens. [Sabol et al., 1983; Sivam et al., 1986; Morris et al., 1988; Angulo, 1992]

II. BIOCHEMICAL EVIDENCE FOR A DOPAMINE-OPIOID INTERACTION

A. DOPAMINE AND OPIOID RECEPTORS SHARE MOLECULAR CHARACTERISTICS IN RECEPTOR FUNCTION AND 2ND MESSENGERS

Molecular cloning of the D_1-D_5 dopamine receptors and the μ and δ opioid receptors reveals that they are all members of the seven transmembrane receptor family that couple to GTP binding (G) proteins. [Surmeier et al., 1993; Chen et al., 1993; Lawrence and Bidlack, 1993] D_1 and D_5 dopamine receptor increases adenylyl cyclase and PKA activity via a G_s protein, whereas D_2 and D_4 dopamine and opioid receptor responses are mediated via G_i or G_o protein inhibition of adenylyl cyclase [Law et al., 1981; Cooper et al., 1986; Childers, 1988; Bunzow et al., 1988; Dearry et al., 1990; Sunahara et al., 1990; Sunahara et al., 1991; Zhou et al., 1990; Van Tol et al., 1991] or opening K^+ channels (see Figure 2 illustrating shared transduction mechanisms).

The intracellular signaling mechanisms for D_3 receptors have not been resolved, but this receptor is unique from the other dopamine receptors in that ligand binding is not sensitive to the guanine nucleotides and does not affect the activity of adenylyl cyclase. [Sokoloff et al., 1990; Sokoloff et al., 1992] Whether the dopamine and opioid receptors coexist in the same cell has not been visualized with anatomical techniques. However, an interaction between dopamine and opioid agonists on adenylyl cyclase activity has been observed, suggesting that these receptors are acting within the same cell to regulate this enzyme. Opioid peptides decrease both basal and dopamine-stimulated adenylyl cyclase activity in the striatum. [Law et al., 1981; Cooper et al., 1986; Childers,

1988; Gentleman et al., 1983] Also, dopamine-stimulated efflux of cyclic AMP, as well as dopamine-stimulated cyclic AMP accumulation, is inhibited by activation of μ-opioid receptors in the rat striatum. [Schoffelmeer et al., 1985; Schoffelmeer et al., 1987; Schoffelmeer et al., 1988; Kelly and Nahorski, 1986]

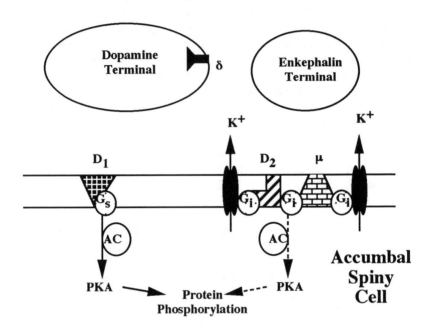

Fig. 2. Illustration of transduction mechanisms for dopamine and opioid receptors which indicates possible points of interaction in the nucleus accumbens. In the membranes of accumbal cells are the D_1 or D_2 family of dopamine receptors as well as a family of opioid receptors (μ and δ), which are linked via either stimulatory (G_s) or inhibitory (G_i or G_o) proteins through adenylyl cyclase, respectively. Adenylyl cyclase produces the 2nd messenger, cyclic AMP, which in turn activates a cyclic AMP-dependent protein kinase that phosphorylates a substrate protein, which may be the D_2 receptor or other receptors. The D_2 or the μ opioid receptors can also regulate the opening of K^+ channels via a G_i or G_o protein. D_2 and δ opioid receptors (represented by funnel symbol) are found on presynaptic dopamine terminals. D_2 receptors increase K^+ conductance, thereby inhibiting dopamine release, and decreasing the activity of tyrosine hydroxylase. The δ opioid receptors act on presynaptic dopamine terminals to stimulate dopamine release.

B. OPIOIDS MAY REGULATE DOPAMINE RECEPTOR ACTION VIA cAMP-DEPENDENT PHOSPHORYLATION

The target for cAMP actions is cAMP-dependent phosphorylation of cyclic AMP-dependent protein kinase (PKA) or other transcription factors. Opioid-induced changes in phosphorylation of brain membranes have been observed in several reports (see, [Childers, 1991] for review). Williams and Clouet [Williams and Clouet, 1982] found that acute administration of opioid agonists altered the phosphorylation of striatal proteins. Nestler and Tallman [Nestler and Tallman, 1988] showed that chronic morphine increased cyclic AMP-dependent protein kinase activity in the VTA and nucleus accumbens. Fleming et al. [Fleming et al., 1992] discovered that opioids inhibited the phosphorylation of Synapsin I and II (synaptic vesicle-associated phosphoproteins) by cyclic AMP-dependent protein kinase in rat striatal membranes. These data suggest that opioids regulate specific phosphoproteins through the inhibition of adenylyl cyclase activity. cAMP-dependent protein phosphorylation regulates the functional state of some receptors [Sibley et al., 1988; Huganir, 1992], including the D_2, GABA, nicotinic and GluR6 receptors as well as some ion channels [Costa et al., 1982; Miles et al., 1987; Browning et al., 1990; Elazar and Fuchs, 1991; Nakayama et al., 1993; Raymond et al., 1993], constituting a potential mechanism for dopamine-opioid interactions.

Repeated administration of either cocaine or morphine produces adaptive alterations in the adenylyl cyclase transduction system in the nucleus accumbens. G_S protein content is elevated, as is the activity of adenylyl cyclase by chronic, but not acute morphine or cocaine treatment. [Nestler, 1992] Associated with these changes is a concomitant activation of cyclic AMP-dependent protein kinase activity. This cascade may be further accentuated by a reduction in G_i protein content. [Striplin and Kalivas, 1993] In the VTA, a similar reduction in G_i protein is produced by both chronic morphine and cocaine. [Nestler et al., 1990] These modifications in the 2nd messenger transduction by both opioids and indirect dopamine agonists support the presence of a common intracellular transduction system for dopamine and endogenous opioids.

C. DOPAMINE MAY REGULATE ENKEPHALIN LEVELS VIA GENE EXPRESSION REGULATORS

Dopamine and opioids may influence gene expression via second messenger-dependent phosphorylation and/or induction of a class of nuclear proteins referred to as transcription factors that bind to specific response elements (e.g. a DNA sequence within the promoter region of genes). These transcription factors control the rate at which the genes are transcribed. The family of related transcription factors that mediate the effects of cAMP and perhaps calcium is called CREB (cAMP response element binding) proteins. [Goodman, 1990; Montminy et al., 1990; Sheng et al., 1991] Protein kinases may also stimulate the production of transcription factors, referred to as immediate early genes, such as the genes that produce the family of proteins, c-fos and c-jun. The dimerization of the cfos/cjun family proteins can activate or repress the transcription of genes by binding to DNA response elements referred to as AP-1 (activated protein) sites, which are present near the preproenkephalin, prodynorphin and tyrosine hydroxylase genes. [Hyman et al., 1988; Cambi et al., 1989; Sheng and Greenberg, 1990; Morgan and Curran, 1991] Acute cocaine and amphetamine administration increase c-fos production in the nucleus accumbens and striatum [Robertson et al., 1992b; Persico et al., 1993], as well as a subregion of the VTA (personal communication,

Dr. Jacqueline McGinty, East Carolina University). Moreover, conditioning to cocaine-paired environment will increase c-fos in specific regions of the brain [Brown et al., 1992] as well as the zinc family of immediate early genes. [Moratalla et al., 1992] Although acute cocaine increases many of the immediate early genes (c-fos, fosB, c-jun, junB and zif268), chronic-non-contingent cocaine or amphetamine treatment with withdrawal prior to the acute administration produces tolerance so that these increases disappear. [Hope et al., 1992; Persico et al., 1993] In contrast, the AP-1 binding site in the nucleus accumbens increases after acute cocaine administration whether or not the rats received a 14 day chronic cocaine treatment with withdrawal 18 hr prior to the acute cocaine administration. These data suggest that persistent changes in gene regulation are occurring during the chronic cocaine administration.   The influence of chronic cocaine on opioid peptide production is suggested by the decreases in preproenkephalin mRNA levels and μ opioid receptor binding as well as increases in dynorphin mRNA levels and κ-opioid receptor binding in cocaine addicts. [Hammer, 1989; Hurd and Herkenham, 1993]

In the VTA evidence for activation of immediate early genes by chronic dopamine or opioid agonists is not clear. Nonetheless, chronic treatment with either cocaine or morphine decreased protein content of G_i and neurofilament proteins, as well as pertussis-toxin ADP-ribosylation of G proteins, while increasing the content of tyrosine hydroxylase. [Nestler et al., 1990; Beitner-Johnson and Nestler, 1991; Striplin and Kalivas, 1992; Beitner-Johnson et al., 1992; Sorg et al., 1993] Neurofilament is a protein thought to be involved in axoplasmic transport and Beitner-Johnson and Nestler [Beitner-Johnson and Nestler, 1993] found that chronic morphine or cocaine decreased the rate of axoplasmic transport in the mesoaccumbens projection. These similarities in the effect of chronic morphine and cocaine on proteins in the VTA are consistent with the existence of a common interaction between opioids, dopamine and genetic expression.

III. ELECTROPHYSIOLOGICAL EVIDENCE FOR DOPAMINE-OPIOID INTERACTIONS

In the VTA, μ-opioid receptor activation in vivo increases the firing frequency of dopamine neurons [Gysling and Wang, 1983; Matthews and German, 1984; Kiyatkin, 1988], whereas δ– and κ-opioid antagonists do not elicit any effects on the dopaminergic or non-dopaminergic neurons. [Johnson and North, 1992] Although systemic administration of the κ-opioid agonist, U50,548, inhibits dopamine neurons in the substantia nigra, pars compacta, the effect was mimicked by U50,548 microinjection into the striatum, indicating the inhibition occurred via the striatonigral pathway. [Walker et al., 1987] The μ-opioid agonists excite the DA neurons in the VTA via inhibition of non-dopaminergic neurons that tonically inhibit these neurons.   Intracellular recordings reveal that this disinhibition occurs via a combination of presynaptic inhibition of tonically-active GABAergic terminals and by postsynaptic inhibition of tonically-active GABA interneurons. [Johnson and North, 1992]

Stimulation of μ-opioid receptors in the VTA primarily inhibits spontaneously active cells in the nucleus accumbens which is blocked by administration of the dopamine receptor antagonist, flupenthixol. [Hakan and Henriksen, 1989] Fimbria-evoked stimulation of accumbal neurons is also inhibited by the morphine injection into the VTA but this stimulation is not blocked by flupenthixol. Morphine iontophoresis directly into the nucleus accumbens also inhibits spontaneous activity but does not affect fimbria-stimulated activity. These data suggest that peripheral morphine can act through dopaminergic and non-dopaminergic projections from the VTA to the nucleus accumbens as well as directly in the accumbens. Furthermore, all three effects produce a decrease in neuronal activity in the nucleus accumbens.

In various brain nuclei, the activation of μ opioid receptors increases K^+ conductance. [Williams et al., 1982; North and Williams, 1985; Uchimura et al., 1986; North et al., 1987] Opioids, particularly μ-opioid selective agonists, hyperpolarize neurons by increasing K^+ conductance, as well as by decreasing a Ca^{+2}-dependent spike and after-hyperpolarization. [Mudge et al., 1979; Duan et al., 1990] The after-hyperpolarization has been suggested to be a Ca^{+2}-dependent K^+ conductance, which can be blocked by phorbol esters at concentrations that activate PKC. [Baraban et al., 1985] Thus, stimulation of PKC by phorbol esters inhibits this action of opioids. Enkephalin has been shown to inhibit Ca^{+2} currents in neuroblastoma cells, [Tsunoo et al., 1986] perhaps through regulation of G_o protein. [Hescheler et al., 1987] δ Opioid agonists also transiently increase Ca^{+2} in these neuroblastoma cells through dihydropyridine-sensitive, voltage-gated Ca^{+2} channels. [Jin et al., 1992] The regulation of Ca^{+2} currents and the inhibition of adenylyl cyclase may be bringing about longer term changes in the neurons, such as a reduction in transcription of certain genes. [North, 1991]

Dynorphin inhibits adenylyl cyclase and reduces calcium currents by activation of a pertussis toxin-sensitive G protein. [Gross et al., 1990] These mechanisms appear to be independent, since dynorphin can reduce calcium currents to an even greater amount in the presence of the catalytic subunit for cAMP-dependent protein kinase. These data suggest that an interaction between the dopamine receptors and the μ- or κ-opioid receptor might also occur at the level of intracellular transduction. Intranigral dynorphin increased dopamine metabolism in the striatum [Tan and Tsou, 1988] while decreasing dopamine release [Reid et al., 1988] suggesting that the increase in dopamine metabolism may be occurring in the absence of enhanced release. [Reid et al., 1990]

In the ventral mesencephalon, dopamine acts through the D_2 receptor family to increase K^+ conductance [Lacey et al., 1987] through a pertussis toxin-sensitive G protein. [Innis and Aghajanian, 1987] D_2 receptors can also activate ATP-stimulated arachidonic acid release when transfected into Chinese hamster ovary cells and this mechanism is blocked by pertussis toxin, although at a different dose from the inhibition of cAMP accumulation. [Felder et al., 1991]

Inhibition of protein kinase C also blocked the augmentation in ATP-stimulated arachidonic acid release, suggesting that the D_2 receptor may regulate other 2nd messenger systems as well. The fact that phorbol esters potentiate dopamine-induced cAMP formation in striatal neurons in culture suggests that cross talk between these two second messenger systems may be relevant to the transduction of dopamine signals. [Schinelli et al., 1994] Voltage-gated Na^+ channels reduce peak current when phosphorylated by either cAMP-dependent protein kinase or protein kinase C, suggesting that a convergent regulation of these two systems might be involved in mediating neuronal signals. [Li et al., 1993] The fact that phorbol esters block the action of opioids suggests that the cross talk with protein kinase C may regulate intracellular processes by a different mechanism for opioids from that produced by dopamine.

IV. BEHAVIORAL EVIDENCE FOR DOPAMINE-OPIOID INTERACTIONS

A. INTERACTION BETWEEN DOPAMINE AND OPIOIDS IN LOCOMOTION

The systemic administration of μ opioid or dopamine agonists elicits an increase in locomotion in rodents. [Randrup and Munkvad, 1967; Creese and Iversen, 1975; Kelly and Iversen, 1976; Babbini and Davis, 1972; Pert and Sivit, 1977] Following systemic administration, the motor stimulant effect of opioids is partly dopamine-dependent. Thus, while the motor response to systemic opioids is blocked by pretreatment with dopamine antagonists [Iwamoto, 1981] and associated with elevated dopamine transmission in the nucleus accumbens [Wood et al., 1980; Kalivas and Stewart, 1991], electrolytic lesions of the nucleus accumbens did not prevent morphine-induced motor activity. [Bunney et al., 1984] The dopamine-dependent and -independent actions of systemic morphine arise from actions in the VTA and nucleus accumbens, respectively. Thus, microinjection of μ- or δ-opioids into the VTA elicits an increase in motor activity that is prevented by dopamine blockers or depletion and is associated with elevated dopamine release in the VTA [Klitenick et al., 1992] and nucleus accumbens. [Kalivas et al., 1983; Vezina and Stewart, 1984; Kelley et al., 1980; Calenco-Choukroun et al., 1991; Kalivas and Duffy, 1990; Dauge et al., 1992; Cador et al., 1989; Leone et al., 1991; Shippenberg, 1992] In contrast, the motor stimulation elicited by opioid microinjection into the nucleus accumbens is not altered by dopamine blockers or dopamine depletions [Pert and Sivit, 1977; Kalivas et al., 1983; Kalivas and Bronson, 1985; Stinus et al., 1985; Maldonado et al., 1990], nor is it associated with enhanced dopamine transmission. [Kalivas et al., 1983] In addition to the VTA and nucleus accumbens, opioids produce an increase in locomotor activity when microinjected into other nuclei of the motive circuit, including the ventral pallidum and mediodorsal thalamus. [Baud et al., 1989; Austin and Kalivas, 1990; Napier, 1992; Klitenick and Kalivas, 1994] In the ventral pallidum, the motor stimulant effect is partly blocked by D_1, but not D_2 antagonists [Napier,

1992; Alesdatter and Kalivas, 1993] while in the mediodorsal thalamus dopamine blockers are ineffective. [Klitenick and Kalivas, 1994] To determine the relative importance of opioid effects in some of the nuclei in the motive circuit, Amalric and Koob [1985] microinjected an opioid antagonist into the VTA or nucleus accumbens and observed that the antagonist more potently blocked the motor stimulation by systemic heroin in the nucleus accumbens than in the VTA.

The motor stimulant effects of pharmacological administration of opioids reflect the actions of endogenous opioids, at least in the VTA and nucleus accumbens. Thus, microinjection of the enkephalinase inhibitors into the VTA produces a motor response that is associated with elevated dopamine transmission in the nucleus accumbens. [Kalivas et al., 1986; Dauge et al., 1992] Likewise, microinjection of inhibitors of enkephalin metabolism into the nucleus accumbens elicits motor activity. [Dauge et al., 1988] In contrast to μ and δ opioid receptor stimulation, neither dynorphin nor κ opioid agonists microinjected into the VTA or nucleus accumbens alter motor activity or DA metabolites in the nucleus accumbens. [Kalivas et al., 1985; Mitchell and Stewart, 1990]

Unlike other nuclei of the motive circuit where opioids elicit a motor response, in the VTA and the ventral pallidum the synaptic mechanisms involved have been partly elucidated. In the ventral pallidum the motor stimulation produced by microinjected Tyr-D-Ala-Gly-NmePhe-Gly-OH (DAMGO) is prevented by stimulating $GABA_A$ receptors with coadministered muscimol, indicating that inhibition of GABA release in the ventral pallidum may play a role. [Austin and Kalivas, 1990] The involvement of inhibition of GABA transmission in the motor response by opioids in the VTA has been more thoroughly characterized. Inhibition of tonically active GABAergic interneurons in the VTA is evidenced by a 40% reduction in extracellular GABA when morphine is administered through a dialysis probe. [Klitenick et al., 1992] A decrease in GABA release by opioids was also observed in nigral slices. [Starr, 1985] As discussed in the previous section, electrophysiological data demonstrate that the disinhibition of the VTA dopamine neurons occurs via a combination of inhibition of tonically active GABAergic terminals and interneurons. [Johnson and North, 1992] Since this reduction in GABA release is associated with an increase in extracellular dopamine and the opioid-induced increase in locomotion is blocked by administering the $GABA_B$ agonist, baclofen, into the VTA [Klitenick et al., 1992], the opioids appear to be influencing the dopamine release indirectly through a GABAergic innervation. This conclusion is supported by the fact that μ and δ opioid agonists do not alter [^{3}H] DA release from ventral mesencephalic neurons in culture. [Smith et al., 1992]

B. INTERACTION BETWEEN OPIOIDS AND DOPAMINE IN REWARD

Opioids, and the indirect dopamine agonists, cocaine and amphetamine, are self-administered. [Werner et al., 1976; Koob and Bloom, 1988] Similar to locomotor activity, a primary anatomical site of action for this effect of opioids and indirect dopamine agonists is the nucleus accumbens. Thus, amphetamine and opioids are self-administered directly into the nucleus accumbens. [Olds, 1982; Hoebel et al., 1983] However, similar to the locomotor stimulation action of intra-accumbens opioids, the reinforcing effect is not dependent upon dopamine transmission. While the systemic self-administration of amphetamine is extinguished by mesoaccumbens dopamine depletions or repeated dopamine antagonist administration, [Lyness et al., 1979; Roberts and Koob, 1982; Ettenberg et al., 1982] the self-administration of morphine is actually augmented. [Stinus et al., 1989] Supporting a critical role for the nucleus accumbens in the systemic self-administration of opioids and amphetamine, intra-accumbens injection of opioid antagonists or dopamine blockers, respectively, increases the rate of self-administration in an effort to overcome the blockade. [Corrigall and Vaccarino, 1988; Vaccarino et al., 1985; Koob et al., 1993] While dopamine blockers were ineffective when microinjected into the VTA, opioid antagonist microinjection into the VTA produced effects similar to intra-accumbens injections, albeit requiring higher doses of antagonist. [Vaccarino et al., 1985] Involvement of the VTA in opioid self-administration is further substantiated by the fact that opioids are self-administered directly into the VTA. [Bozarth and Wise, 1981] Furthermore, reward following intra-VTA opioids is abolished by mesoaccumbens dopamine lesions. [Phillips et al., 1983] Thus, similar to locomotor stimulation, the rewarding effects of opioids in the VTA are dependent upon activation of mesoaccumbens dopamine transmission. Likewise, following systemic opioids, this action in the VTA is secondary to the dopamine-independent effect in the nucleus accumbens.

C.   DOPAMINE-OPIOID   INTERACTIONS   IN   MESOLIMBIC CIRCUITRY INTERDEPENDENCE

The data outlined above demonstrate anatomical, pharmacological and physiological evidence for an interaction between opioids and the mesolimbic dopamine transmission. As indicated in the anatomical section, the regulation of locomotion involves an interdependence between the dopaminergic neurons in the VTA, the GABAergic and enkephalinergic neurons in the nucleus accumbens and the GABAergic neurons in the ventral pallidum (see Fig. 1). This interdependence was initially revealed by the fact that locomotion induced by microinjection of DAMGO into the ventral pallidum was blocked by a dopaminergic receptor antagonist in the nucleus accumbens. [Austin and Kalivas, 1991] These data suggested that the regulation of locomotion did not involve a linear route from the dopaminergic neurons of the VTA through the accumbens to the ventral pallidum and out to motor nuclei, but rather depended on an intact circuitry interconnecting these three areas. The concept that locomotion induced in the ventral pallidum of the mesolimbic locomotor circuit depends on a cycling of information back through the VTA to the nucleus accumbens suggests that the opioid-dopamine interaction in the VTA and accumbens might be involved.

Recent studies of the anatomical and pharmacological details of the mesolimbic locomotor circuitry reveal that specific compartments of the nucleus accumbens and ventral pallidum exist and may be important in the regulation of locomotion (for review, see [Zahm and Brog, 1992; Groenewegen et al., 1993; Kalivas et al., 1993a]). The nucleus accumbens is compartmentalized into a core and shell which topographically interconnects with a dorsolateral and ventromedial ventral pallidum, respectively. The accumbens shell and the ventromedial pallidum interconnect with the VTA; whereas the accumbens core and dorsolateral ventral pallidum project to the medial substantia nigra compacta. The output from the ventral pallidum through the mediodorsal thalamus arises primarily from the ventral parts and projects mainly to a region of the prefrontal cortex that returns to the accumbens core (see [Deutch et al., 1993]). This circuitry suggests that the regulation of locomotion within the circuit may be topographically organized into distinct outputs via the accumbens core and dorsolateral ventral pallidum on one hand and the accumbens shell, ventral parts of the ventral pallidum and VTA on the other.

By blocking various output pathways from the ventral pallidum while inducing locomotion from different compartments within the mesolimbic locomotor circuits, the relationship between these various compartments can be evaluated. Locomotion induced by a μ opioid agonist, DAMGO, in the accumbens core is blocked by a local anesthetic, procaine, in the mediodorsal thalamus, but not by hyperpolarizing dopamine neurons in the VTA with microinjection of the GABA$_B$ agonist, baclofen. In contrast, locomotion induced by DAMGO in the accumbens shell is not blocked by procaine in the mediodorsal thalamus but by baclofen in the VTA. [Churchill and Kalivas, 1993] These results suggest that the interconnections between the accumbens shell and VTA are essential for opioid-induced locomotion from the shell and that the output from this circuitry does not depend on the mediodorsal thalamus. However, the opioid-induced locomotion from the accumbens core does depend on neuronal activity in the mediodorsal thalamus. These data suggest that a topographic flow of information resulting from opioid-induced alterations in neuronal activity in the motive circuit does indeed exist. However, it is not clear how this contributes to the behavioral response nor if there exists a differential interaction between opioids and dopamine depending upon the subregions involved in the opioid-mediated responses.

D. EFFECT OF MESOACCUMBENS DOPAMINE LESIONS ON OPIOID-INDUCED BEHAVIORS

The literature outlined above clearly indicates that the effect of opioids in the nucleus accumbens is not dependent upon intact dopamine innervation. Thus, 6-OHDA lesions of the mesoaccumbens dopamine projection inhibit neither opioid-induced locomotion nor reward. In fact, both of these behavioral effects of opioids are augmented by chronically damaging dopamine transmission either with lesions or repeated administration of dopamine blockers. [Stinus et al., 1985; Kalivas and Bronson, 1985; Stinus et al., 1989]

The μ1 opioid receptor subtype plays an important role in the augmented locomotor response observed following 6-OHDA lesions of the nucleus accumbens. Microinjection of the μ opioid receptor agonist, DAMGO, into the NA produces a moderate increase in motor activity which is dramatically enhanced following dopamine lesions. [Churchill and Kalivas, 1992]   In contrast, microinjection of the δ opioid agonist,   [D-penicillamine^{2,5}]-enkephalin (DPDPE), into the nucleus accumbens elicits a robust motor stimulant effect [Dauge et al., 1988] which is not augmented following dopamine depletion. [Churchill and Kalivas, 1992] Dopamine lesion-induced augmentation of DAMGO is not caused by a measurable increase in the density of μ opioid receptors in the nucleus accumbens, but may result from modifications in the postsynaptic transduction mechanisms. [Churchill and Kalivas, 1992] This is in contrast to the behavioral sensitization to dopamine following accumbal dopamine lesions which are associated with an enhanced D_2 receptor density [Savasta et al., 1988], but not D_1-stimulated adenylate cyclase. [Krueger et al., 1976]   Considering that opioids and dopamine receptors impinge, in part, upon identical intracellular transduction substrates it may be that the upregulation for the stimulation of both dopamine and μ opioid receptors results from modifications of the same transduction processes by presynaptic dopamine depletions.   While, this hypothesis remains to be fully evaluated for dopamine depletions, excessive stimulation of either dopamine or opioid receptors by repeated administration of amphetamine or morphine, respectively, produces increases in cAMP-dependent phosphorylation for specific proteins in the nucleus accumbens. [Guitart and Nestler, 1990] Preliminary data suggest that a similar increase in cAMP-dependent phosphorylation for a 17 kDa protein, perhaps a myelin basic protein, occurs 10 days after dopamine depletion in the nucleus accumbens.

The capacity of endogenously released opioids to initiate motor activity is also augmented after the dopamine lesions. Microinjection of the inhibitor of enkephalin degradation, kelatorphan, into the nucleus accumbens produces a dose-dependent locomtor response which is enhanced in dopamine-lesioned versus intact animals. [Kalivas et al., 1993a]   A μ1 selective antagonist, naloxonazine, as well as a δ selective antagonist, naltrindole, partially blocks the augmentation in locomotion produced by kelatorphan, suggesting that both opioid receptor subtypes are involved in the augmented response to the endogenous opioids in animals sustaining dopamine lesions of the nucleus accumbens.

The augmentation of opioid transmission by 6-OHDA lesions is reflected in a functional alteration in the motive circuit shown in Figure 1.   Many laboratories have shown that locomotor activity can be stimulated by inhibition of GABA$_A$ receptors following microinjection of picrotoxin into the ventral pallidum. [Swerdlow and Koob, 1987; Mogenson and Wu, 1988; Austin and Kalivas, 1990] This response is not blocked by pretreatment with naloxone, indicating that opioid receptor stimulation is not required. However, in animals sustaining mesoaccumbens dopamine depletions picrotoxin-induced locomotion is blocked by systemic naloxone administration. [Churchill et al., 1992] More recently, this blockade in lesioned animals was replicated by microinjection of naltrexone methobromide directly into the nucleus accumbens. [Kalivas et al., 1993a] This demonstrates that after mesoaccumbens dopamine depletion opioid transmission in the nucleus accumbens assumes a more critical role in the expression of locomotor activity than in animals with normal dopamine transmission.

E. DOPAMINE-OPIOID INTERACTIONS IN NEUROPSYCHIATRY

Alterations or damage to components of the motive circuit depicted in Figure 1 have been proposed as pathological substrates for certain neuropsychiatric disorders. The literature evaluated above clearly implicates nucleus accumbens as a site mediating abuse of both opioids and amphetamine-like psychostimulants. In this regard, opioid and dopamine influences in preclinical studies were found to be partly interdependent. Thus, it may be possible to modulate opioid transmission or dopamine transmission to affect substance abuse patterns in humans. To this end, it was recently shown that buprenorphine, a partial opioid agonist, blocks self-administration of cocaine in monkeys and reduces the cocaine present in addicts trying to reduce their intake. [Jasinski et al., 1989; Mello et al., 1989; Schottenfeld et al., 1993]

Schizophrenia has long been linked to alterations in dopamine transmission in the nucleus accumbens and prefrontal cortex. [Davis et al., 1991; Csernansky et al., 1991; Lipska and Weinberger, 1993] A model of dopamine-dependent psychosis is amphetamine-induced psychosis. Repeated amphetamine administration produces paranoid behavior in humans and sensitization of locomotion in rodents. [Kokkinidis and Anisman, 1980; Robinson and Becker, 1986] The sensitized behavioral response to amphetamine is mediated in part by enhanced dopamine release in the nucleus accumbens. [Robinson and Becker, 1986; Kalivas and Stewart, 1991] Repeated administration of morphine in rodents also produces behavioral sensitization [Kalivas and Duffy, 1987] that demonstrates cross-sensitization with amphetamine-like psychostimulants. [Vezina and Stewart, 1990] Moreover, similar to amphetamine, sensitization to morphine is associated with augmented dopamine release in the nucleus accumbens. [Kalivas and Stewart, 1991] Inasmuch as sensitization is a model for the development of schizophrenia, it is possible that opioid antagonists may be useful alone or in combination with dopamine blockers in treating some symptoms associated with schizophrenia. Treatment of schizophrenics with the opioid antagonist, naloxone, significantly reduced symptoms 2-7 hr after administration. [Emrich et al., 1977; Berger et al., 1981] The delayed effect may be due to a compensatory release of endogenous opioids [Volavka et al., 1980], suggesting that an agonistic action might be responsible for the antipsychotic effects. [Emrich and Schmauss, 1977] Treatment of severely ill, chronic schizophrenics with methadone added to a neuroleptic therapy seemed to be effective. [Brizer et al., 1985] Recently, buprenorphine was used to treat ten neuroleptic-free schizophrenics suffering from frequent hallucinations and severe formal thought disorders. [Schmauss et al., 1987]   These patients appeared to improve dramatically for about 4 hr; whereas patients with residual schizophrenia were not improved by this treatment.   These results suggest that the endogenous opioids may play a role in the hallucinations and delusions observed in schizophrenics. However, the effects observed to date are mild, indicating at best a permissive or modulatory role, not a necessary role for opioid transmission in schizophrenia.

CONCLUSIONS

We can conclude that there exist interactions between opioid and dopamine transmission in the motive circuit shown in Figure 1. These interactions do not occur at the same molecular binding site but are by virtue of cellular transduction mechanisms. Opioid interactions with dopamine transmission have been shown at three loci, 1) presynaptic alterations in dopamine release, 2) indirect stimulation of dopamine neuronal activity and 3) postsynaptic interactions. Modulation of GABA transmission by opioids mediates the action on dopamine cells and perhaps the pre- and postsynaptic effects in dopamine terminal fields as well.

The motive circuit is a critical gate in translating motivationally relevant stimuli into adaptive behavioral responses. Both opioids and dopamine are clearly implicated as modulators in this gating function, as evidenced by their rewarding properties and their ability to produce behavioral sensitization. Furthermore, inasmuch as many psychopathologies involve maladaptive behavioral responses to environmental stimuli the role of dopamine and opioids may be clinically relevant. Unfortunately, to date, a role for an interaction between dopamine and opioids in neuropsychiatric disorders is not evidenced by clearly successful therapies utilizing opioid agonists or antagonists. However, the synaptic organization permitting opioid and dopamine transmission to interact and modulate the status of the motive circuit remains poorly described. As the arrangement and relative roles that opioids and dopamine have in regulating the motive circuit are elucidated the potential of this interaction for novel psychotherapies will be clarified.

REFERENCES

Alesdatter, J.E. and Kalivas, P.W., Blockade of opioid induced motor activity by dopamine blockers in the ventral pallidum, *Behav. Pharmacol.*, 4, 645, 1993.

Amalric, M. and Koob, G.F., Low doses of methylnaloxonium in the nucleus accumbens antagonize hyperactivity induced by heroin in the rat, *Pharmacol. Biochem. Behav.*, 23, 411, 1985.

Angulo, J.A., Involvement of dopamine D1 and D2 receptors in the regulation of proenkephalin mRNA abundance in the striatum and accumbens of the rat brain, *J. Neurochem.*, 58, 1104, 1992.

Ariano, M.A., Stromski, C.J., Smyk-Randall, E.M. and Sibley, D.R., D_2 dopamine receptor localization on striatonigral neurons, *Neurosci. Lett.*, 144, 215, 1992.

Austin, M.C. and Kalivas, P.W., Enkephalinergic and GABAergic modulation of motor activity in the ventral pallidum, *J. Pharmacol. Exp. Ther.*, 252, 1370, 1990.

Austin, M.C. and Kalivas, P.W., Dopaminergic involvement in locomotion elicited from the ventral pallidum/substantia innominata, *Brain Res.*, 542, 123, 1991.

Babbini, M. and Davis, W.M., Time-dose relationships for locomotor activity effects of morphine after acute or repeated treatment, *Brit. J. Pharmacol.*, 46, 213, 1972.

Baraban, J.M., Snyder, S.H. and Alger, B.E., Protein kinase C regulates ionic conductance in hippocampal pyramidal neurons: electrophysiological effects of phorbol esters, *Proc. Natl. Acad. Sci. USA*, 82, 2538, 1985.

Baud, P., Mayo, W., le Moal, M. and Simon, H., Locomotor hyperactivity in the rat after infusion of muscimol and [D-Ala^2]Met-enkephalin into the nucleus basalis magnocellularis. Possible interaction with cortical cholinergic projections, *Brain Res.*, 452, 203, 1989.

Bayer, V.E. and Pickel, V.M., GABA-labeled terminals form proportionally more synapses with dopaminergic neurons having low densities of tyrosine hydroxylase immunoreactivity in rat ventral tegmental area, *Brain Res.*, 559, 44, 1991.

Beitner-Johnson, D. and Nestler, E.J., Morphine and cocaine exert common chronic actions on tyrosine hydroxylase in dopaminergic brain reward regions, *J. Neurochem.*, 57, 344, 1991.

Beitner-Johnson, D., Guitart, X. and Nestler, E.J., Common intracellular actions of chronic morphine and cocaine in dopaminergic brain reward regions, in *The Neurobiology of Drug and Alcohol Addiction*, Kalivas, P.W. and Samson, H.H., Eds. Ann. N.Y. Acad. Sci., New York, NY, 1992, p. 70-87.

Beitner-Johnson, D.J. and Nestler, E.J., Chronic morphine impairs axoplasmic transport in the rat mesolimbic dopamine system, *Neuroreport*, 5, 57, 1993.

Berger, P.A., Watson, S.J., Akil, H. and Barchas, J.D., The effects of naloxone in chronic schizophrenia, *Am. J. Psychiat.*, 38, 913, 1981.

Besson, M.J., Graybiel, A.M. and Quinn, B., Co-expression of neuropeptides in the cat's striatum: an immunohistochemical study of substance P, dynorphin B and enkephalin, *Neuroscience*, 39, 33, 1990.

Bozarth, M.A. and Wise, R.A., Intracranial self-administration of morphine into the ventral tegmental area in rats, *Life Sci.*, 29, 551, 1981.

Brizer, D.A., Hartmann, N., Sweeney, J. and Millman, R.B., Effect of methadone plus neuroleptics on treatment-resistant chronic paranoid schizophrenia, *Am. J. Psychiat.*, 142, 1106, 1985.

Brown, E.E., Robertson, G.S. and Fibiger, H.C., Evidence for conditional neuronal activation following exposure to a cocaine-paired environment: role of forebrain limbic structures, *J. Neurosci.*, 12, 4112, 1992.

Browning, M.D., Bureau, M., Dudek, E.M. and Olsen, R.W., Protein kinase C and cAMP-dependent protein kinase phosphorylate the β subunit of the purified γ-aminobutyric acid A receptor, *Proc. Natl. Acad. Sci. USA*, 87, 1315, 1990.

Bunney, W.C., Massar, V.J. and Pert, A., Chronic morphine-induced hyperactivity in rats is altered by nucleus accumbens and ventral tegmental lesions, *Psychopharmacol.*, 82, 318, 1984.

Bunzow, J.R., Van Tol, H.H.M., Grandy, D.K., Albert, P., Salon, J., Christie, M., Machida, C.A., Neve, K.A. and Civelli, O., Cloning and expression of a rat D_2 dopamine receptor cDNA, *Nature (London)*, 336, 783, 1988.

Cador, M., Rivet, J.-M., Kelley, J.-M., LeMoal, M. and Stinus, L., Substance P, neurotensin and enkephalin injections into the ventral tegmental area: a comparative study on dopamine turnover in several forebrain structures, *Brain Research*, 486, 357, 1989.

Calenco-Choukroun, G., Dauge, V., Gacel, G., Feger, J. and Roques, B.P., Opioid δ agonists and endogenous enkephalins induce different emotional reactivity than μ agonists after injection in the rat ventral tegmental area, *Psychopharmacol.*, 103, 493, 1991.

Cambi, F., Fung, B. and Chikaraishi, D., 5' flanking DNA sequences direct cell-specific expression of rat tyrosine hydroxylase, *J. Neurochem.*, 53, 1656, 1989.

Cenci, M.A., Campbell, K. and Bjorklund, A., Neuropeptide messenger RNA expression in the 6-hydroxydopamine-lesioned rat striatum reinnervated by fetal dopaminergic transplants: differential effects of the grafts on preproenkephalin, preprotachykinin and prodynorphin messenger RNA levels, *Neuroscience*, 57, 275, 1993.

Chen, Y., Mestek, A., Liu, J., Hurley, J.A. and Yu, L., Molecular cloning and functional expression of a μ-opioid receptor from rat brain, *Mol. Pharmacol.*, 44, 8, 1993.

Childers, S.R., Opiate-inhibited adenylate cyclase in rat brain membranes depleted of G_S-stimulated adenylate cyclase, *J. Neurochem.*, 50, 543, 1988.

Childers, S.R., Opioid receptor-coupled second messenger systems, *Life Sci.*, 48, 1991, 1991.

Churchill, L., Austin, M.C. and Kalivas, P.W., Dopamine and endogenous opioid regulation of picrotoxin-induced locomotion in the ventral pallidum after dopamine depletion in the nucleus accumbens, *Psychopharmacol.*, 108, 141, 1992.

Churchill, L. and Kalivas, P.W., Dopamine depletion produces augmented behavioral responses to a mu, but not a delta-opioid agonist in the nucleus accumbens: Lack of a role for receptor upregulation, *Synapse*, 11, 47, 1992.

Churchill, L. and Kalivas, P.W., Locomotion stimulated from either the nucleus accumbens or the ventral pallidum is mainly regulated by the ventral tegmental area rather than the mediodorsal thalamus, *Soc. Neurosci. Abst.*, 19, 148, 1993. (Abstract)

Churchill, L. and Kalivas, P.W., A topographically organized GABAergic projection from the ventral pallidum to the nucleus accumbens in the rat, *J. Comp. Neurol.*, 345, 579, 1994.

Cooper, D.M.F., Londos, C., Gill, D.L. and Rodbell, M., Opiate-receptor-mediated inhibition of adenylate cyclase in rat striatal plasma membranes, *J. Neurochem.*, 38, 1164, 1986.

Corrigall, W.A. and Vaccarino, F.J., Antagonist treatment in nucleus accumbens or periaqueductal grey affects heroin self-administration, *Pharmacol. Biochem. Behav.*, 30, 443, 1988.

Costa, M.R., Casnellie, J.E. and Catterall, W.S., Selective phosphorylation of the alpha subunit of the sodium channel by cAMP-dependent protein kinase, *J. Biol. Chem.*, 257, 7918, 1982.

Creese, I. and Iversen, S.D., The pharmacological and anatomical substrates of the amphetamine response in the rat, *Brain Res.*, 83, 419, 1975.

Csernansky, J.G., Murphy, G.M. and Faustman, W.O., Limbic/mesolimbic connections and the pathogenesis of schizophrenia, *Biol. Psychiat.*, 30, 383, 1991.

Dauge, V., Rossignol, P. and Roques, B.P., Comparison of the behavioral effects induced by administration in rat nucleus accumbens or nucleus caudatus of selective μ and δ opioid peptides or kelatorphan, an inhibitor of enkephalin-degrading enzymes, *Psychopharmacol.*, 96, 343, 1988.

Dauge, V., Kalivas, P.W., Duffy, T. and Roques, B.P., Effect of inhibiting enkephalin catabolism in the VTA on motor activity and extracellular dopamine, *Brain Res.*, 599, 209, 1992.

Davis, K.L., Kahn, R.S., Ko, G. and Davidson, M., Dopamine in schizophrenia: a review and reconceptualization, *Am. J. Psychiat.*, 148, 1474, 1991.

Dearry, A., Gingrich, J.A., Falardeau, P., Fremeau, R.T., Jr., Bates, M.D. and Caron, M., Molecular cloning and expression of the gene for a human D_1 dopamine receptor, *Nature*, 347, 72, 1990.

Deutch, A.Y., Bourdelais, A.J. and Zahm, D.S., The nucleus accumbens core and shell: accumbal compartments and their functional attributes, in *Limbic Motor Circuits and Neuropsychiatry*, Kalivas, P.W. and Barnes, C.D., Eds. CRC Press, Boca Raton, 1993, p. 45-88.

Duan, S.M., Shimizu, N., Fukuda, A., Hori, T. and Oomura, Y., Hyperpolarizing action of enkephalin on neurons in the dorsal motor nucleus of the vagus, in vitro, *Brain Res. Bull.*, 25, 551, 1990.

Elazar, Z. and Fuchs, S., Phosphorylation by cyclic AMP-dependent protein kinase modulates agonist binding to the D_2 dopamine receptor, *J. Neurochem.*, 56, 75, 1991.

Emrich, H.M., Cording, C., Piree, S., Kolling, W., von Zerssen, D. and Herz, A., Indication of an antipsychotic action of the opiate antagonist naloxone, *Pharmacopsychiat.*, 10, 269, 1977.

Emrich, H.M. and Schmauss, C., Psychiatric aspects of opioid research, in *Neurobiology of Opioids*, Almeida, O.F.X. and Shippenberg, T.S., Eds. Springer-Verlag, New York, 1977, p. 363-367.

Ettenberg, A., Pettit, H.O., Bloom, F.E. and Koob, G.F., Heroin and cocaine intravenous self-administration in rats: mediation by separate neural systems, *Psychopharmacol.*, 78, 204, 1982.

Fallon, J.H. and Moore, R.Y., Catecholamine innervation of basal forebrain. IV. Topography of the dopamine projection to the basal forebrain and striatum, *J. Comp. Neurol.*, 180, 545, 1978.

Felder, C.C., Williams, H.L. and Axelrod, J., A transduction pathway associated with receptors coupled to the inhibitory guanine nucleotide binding protein G_i that amplifies ATP-mediated arachadonic acid release, *Proc. Natl. Acad. Sci. USA*, 88, 6477, 1991.

Fleming, L.M., Ponjee, G., and Childers, S.R., Inhibition of protein phosphorylation by opioid-inhibited adenylyl cyclase in rat brain membranes, *J. Pharmacol. Exp. Ther.*, 260, 1416, 1992.

Gentleman, S., Parenti, M., Neff, N.H. and Pert, C.B., Inhibition of dopamine-activated adenylate cyclase and dopamine binding by opiate receptors in rat striatum, *Cell. Mol. Neurobiol.*, 3, 17, 1983.

Gerfen, C.R., Engbar, T.M., Mahan, L.C., Susel, Z., Chase, T.N., Monsma, F.J., Jr. and Sibley, D.R., D_1 and D_2 dopamine receptor-regulated gene expression of striatonigral and striatopallidal neurons, *Science*, 250, 1429, 1990.

Gerfen, C.R., McGinty, J.F. and Young, W.S., III, Dopamine differentially regulates dynorphin, substance P, and enkephalin expression in striatal neurons: In situ hybridization histochemical analysis, *J. Neurosci.*, 11, 1016, 1991.

Goodman, R.H., Regulation of neuropeptide gene expression, *Annu. Rev. Neurosci.*, 13, 111, 1990.

Groenewegen, H.J., Berendse, H.W. and Haber, S.N., Organization of the output of the ventral striatopallidal system in the rat: ventral pallidal efferents, *Neuroscience*, 57, 113, 1993.

Gross, R.A., Moises, H.C., Uhler, M.D. and MacDonald, R.L., Dynorphin A and cAMP-dependent protein kinase independently regulate neuronal calcium currents, *Proc. Natl. Acad. Sci. USA*, 87, 7025, 1990.

Guitart, X. and Nestler, E.J., Identification of MARPP (14-20), morphine- and cyclic AMP-regulated phosphoproteins of 14-20 kDa, as myelin basic proteins: evidence for their acute and chronic regulation by morphine in rat brain, *Brain Res.*, 516, 57, 1990.

77

Gysling, K. and Wang, R.Y., Morphine-induced activation of A10 dopamine neurons in the rat brain, *Brain Research*, 277, 119, 1983.

Haber, S.N. and Nauta, W.J.H., Ramifications of the globus pallidus in the rat as indicated by patterns of immunohistochemistry, *Neuroscience*, 9, 245, 1983.

Hakan, R.L. and Henriksen, S.J., Opiate influences on nucleus accumbens neuronal electrophysiology: dopamine and non-dopamine mechanisms, *J. Neurosci.*, 9, 3538, 1989.

Hammer, R.P., Jr., Cocaine alters opiate receptor binding in critical brain reward regions, *Synapse*, 3, 55, 1989.

Harrison, M.B., Wiley, R.G. and Wooten, G.F., Changes in D_2 but not D_1 receptor binding in the striatum following a selective lesion of striatopallidal neurons, *Brain Res.*, 590, 305, 1992.

Hescheler, J., Rosenthal, W., Trautwein, W. and Schultz, G., The GTP-binding protein, G_o, regulates neuronal calcium channels, *Nature*, 325, 445, 1987.

Hoebel, B.G., Monaco, A.P., Hernandez, L., Aulisi, E.F., Stanley, B.J. and Lenard, L., Self-injection of amphetamine directly into the brain, *Psychopharmacol.*, 81, 158, 1983.

Hope, B., Kosofsky, B., Hyman, S.E. and Nestler, E.J., Regulation of immediate early gene expression and AP-1 binding in the rat nucleus accumbens by chronic cocaine, *Proc. Natl. Acad. Sci. (USA)*, 89, 5764, 1992.

Huganir, R.L., Regulation of the nicotinic acetylcholine receptor by serine and tyrosine protein kinases, *Adv. Exp. Med. Biol.*, 287, 279, 1992.

Hurd, Y.L. and Herkenham, M., Molecular alterations in the neostriatum of human cocaine addicts, *Synapse*, 13, 357, 1993.

Hyman, W.E., Comb, M., Lin, Y.-S., Pearlberg, J., Green, M.R. and Goodman, H.M., A common trans-acting factor is involved in transcriptional regulation of neurotransmitter genes by cyclic AMP, *Mol. Cell Biol.*, 8, 4225, 1988.

Ingham, C.A., Hood, S.H. and Arbuthnott, G.W., A light and electron microscopical study of enkephalin-immunoreactive structures in the rat neostriatum after removal of the nigrostriatal dopaminergic pathway, *Neuroscience*, 42, 715, 1991.

Innis, R.B. and Aghajanian, G.K., Pertussis toxin blocks autoreceptor-mediated inhibition of dopaminergic neurons in rat substantia nigra, *Brain Research*, 411, 139, 1987.

Iwamoto, E.T., Locomotor activity and antinociception after putative mu, kappa and sigma opioid receptor agonists in the rat: Influence of dopaminergic agonists and antagonists, *J. Pharmacol. Exp. Ther.*, 217, 451, 1981.

Izzo, P.N., Graybiel, A.M. and Bolam, J.P., Characterization of substance P- and [met]enkephalin-immunoreactive neurons in the caudate nucleus of cat and ferret by a single section Golgi procedure, *Neuroscience*, 20, 577, 1987.

Jasinski, D.R., Fudala, P.J. and Johnson, R.E., Sublingual versus subcutaneous buprenorphine in opiate abusers, *Clin. Pharmacol. Ther.*, 45, 513, 1989.

Jin, W., Lee, N.M., Loh, H.H. and Thayer, S.A., Dual excitatory and inhibitory effects of opioids on intracellular calcium in neuroblastoma x glioma hybrid NG 108-15 cells, *Mol. Pharmacol.*, 42, 1083, 1992.

Johnson, S.W. and North, R.A., Opioids excite dopamine neurons by hyperpolarization of local interneurons, *J. Neurosci.*, 12, 483, 1992.

Joyce, J.N., Janowsky, A. and Neve, K.A., Characterization and distribution of [^{125}I]epidepride binding to dopamine D$_2$ receptors in basal ganglia and cortex of human brain, *J. Pharmacol. Exp. Ther.*, 257, 1253, 1991.

Kalivas, P.W., Widerlov, E., Stanley, D., Breese, G.R. and Prange, A.J., Jr., Enkephalin action on the mesolimbic dopamine system: A dopamine-dependent and a dopamine-independent increase in locomotor activity, *J. Pharmacol. Exp. Ther.*, 227, 229, 1983.

Kalivas, P.W. and Bronson, M., Mesolimbic dopamine lesions produce an augmented response to enkephalin, *Neuropharmacol.*, 24, 931, 1985.

Kalivas, P.W., Taylor, S. and Miller, J.S., Sensitization to repeated enkephalin administration into the ventral tegmental area of the rat. I. Behavioral characterization, *J. Pharmacol. Exp. Ther.*, 235, 537, 1985.

Kalivas, P.W., Richardson-Carlson, R. and Van Orden, G., Cross-sensitization between foot shock stress and enkephalin-induced motor activity, *Biol. Psychiat.*, 21, 939, 1986.

Kalivas, P.W. and Duffy, P., Sensitization to repeated morphine injection in the rat: Possible involvement of A10 dopamine neurons, *J. Pharmacol. Exp. Ther.*, 241, 204, 1987.

Kalivas, P.W. and Duffy, P., Effect of acute and daily neurotensin and enkephalin treatments on extracellular dopamine in the nucleus accumbens, *J. Neurosci.*, 10, 2940, 1990.

Kalivas, P.W. and Stewart, J., Dopamine transmission in the initiation and expression of drug- and stress-induced sensitization of motor activity, *Brain Res. Rev.*, 16, 223, 1991.

Kalivas, P.W., Churchill, L. and Klitenick, M.A., The circuitry mediating the translation of motivational stimuli into adaptive motor responses, in *Limbic Motor Circuits and Neuropsychiatry*, Kalivas, P.W. and Barnes, C.D., Eds. CRC Press, Boca Raton, 1993a, p. 237-287.

Kalivas, P.W., Churchill, L. and Klitenick, M.A., GABA and enkephalin projection from the nucleus accumbens and ventral pallidum to VTA, *Neuroscience*, 57, 1047, 1993b.

Kelley, A.E., Stinus, L. and Iversen, S.D., Interactions between D-Ala-Met-enkephalin, A10 dopaminergic neurons, and spontaneous behavior in the rat, *Behav. Brain Res.*, 1, 3, 1980.

Kelly, E. and Nahorski, S.R., Specific inhibition of dopamine-D1-mediated cyclic AMP formation by dopamine D-2, muscarinic cholinergic and opiate receptor stimulation in rat striatal slices, *J. Neurochem.*, 47, 1512, 1986.

Kelly, P.H. and Iversen, S.D., Selective 6-OHDA-induced destruction of mesolimbic dopamine neurons: Abolition of psychostimulant-induced locomotor activity in rats, *Eur. J. Pharmacol.*, 40, 45, 1976.

Kiyatkin, E.A., Morphine-induced modification of the functional properties of ventral tegmental area neurons in conscious rat, *Intern. J. Neuroscience,* 41, 57, 1988.

Klitenick, M.A., DeWitte, P. and Kalivas, P.W., Regulation of somatodendritic dopamine release in the ventral tegmental area by opioids and GABA: an *in vivo* microdialysis study, *J. Neurosci.,* 12, 2623, 1992.

Klitenick, M.A. and Kalivas, P.W., Behavioral and neurochemical studies of opioid effects in the pedunculopontine nucleus and mediodorsal thalamus, *J. Pharmacol. Exp. Ther.,* 269, 437, 1994.

Kokkinidis, L. and Anisman, H., Amphetamine models of paranoid schizophrenia: an overview and elaboration of animal experimentation, *Psychol. Bull.,* 88, 551, 1980.

Koob, G.F. and Bloom, F.E., Cellular and molecular mechanisms of drug dependence, *Science,* 242, 715, 1988.

Koob, G.F., Robledo, P., Markou, A. and Caine, S.B., The mescorticolimbic circuit in drug dependence and reward - A role for the extended amygdala? in *Limbic Motor Circuits and Neuropsychiatry,* Kalivas, P.W. and Barnes, C.D., Eds. CRC Press, Boca Raton, FL, 1993, p. 289-310.

Kosterlitz, H.W., Opioid receptor subtypes: past, present and future, in *Neurobiology of Opioids,* Almeida, O.F.X. and Shippenberg, T.S., Eds. Springer-Verlag, New York, 1991, p. 3-9.

Krueger, B.K., Forn, J., Walters, J.R., Roth, R.H. and Greengard, P., Stimulation by dopamine of adenosine cyclic 3',5'-monophosphate formation in rat caudatus nucleus: Effect of lesions of the nigroneostriatal pathway, *Mol. Pharmacol.,* 12, 639, 1976.

Lacey, M.G., Mercuri, N.B. and North, R.A., Dopamine acts at D_2 receptors to increase potassium conductance in neurones of the rat substantia nigra, *J. Physiol. (Lond),* 392, 397, 1987.

Law, P.Y., Wu, J., Koehler, J.E. and Loh, H.H., Demonstration and characterization of opiate inhibition of the striatal adenylate cyclase, *J. Neurochem.,* 36, 1834, 1981.

Lawrence, D.M.P. and Bidlack, J.M., The kappa opioid receptor expressed on the mouse R1.1thymoma cell line is coupled to adenylyl cyclase through a pertussis toxin-sensitive guanine nucleotide-binding regulatory protein, *J. Pharmacol. Exp. Ther.,* 266, 1678, 1993.

Le Moine, C., Normand, E. and Block, B., Phenotypical characterization of the rat striatal neurons expressing the D_1 dopamine receptor gene, *Proc. Natl. Acad. Sci. USA,* 88, 4205, 1991.

LeMoal, M. and Simon, H., Mesocorticolimbic dopaminergic network: functional and regulatory roles, *Physiol. Rev.,* 71, 155, 1991.

Leone, P., Pocock, D. and Wise, R.A., Morphine-dopamine interaction: Ventral tegmental morphine increases nucleus accumbens dopamine release, *Pharmacol. Biochem. Behav.,* 39, 469, 1991.

Levesque, D., Diaz, J., Pilon, C., Martes, M.-P., Giros, B., Souil, E., Schott, D., Morgat, J.-L., Schwartz, J.-C. and Sokoloff, P., Identification, characterization, and localization of the dopamine D_3 receptor in rat brain using 7- [^{3}H]hydroxy-N,N-di-n-propyl-2-aminotetralin, *Proc. Natl. Acad. Sci. USA*, 89, 8155, 1992.

Li, M., West, J.W., Numann, R., Murphy, B.J., Scheuer, T. and Catterall, W.A., Convergent regulation of sodium channels by protein kinase C and cAMP-dependent protein kinase, *Science*, 261, 1439, 1993.

Liang, C.-L., Kozlowski, G.P. and German, D.C., Leucine[5]-enkephalin afferents to midbrain dopaminergic neurons: light and electron microscopic examination, *J. Comp. Neurol.*, 332, 269, 1993.

Lipska, B.K. and Weinberger, D.R., Cortical regulation of the mesolimbic dopamine system: implications for schizophrenia, in *Limbic Motor Circuits and Neuropsychiatry*, Kalivas, P.W. and Barnes, C.D., Eds. CRC Press, Boca Raton, FL, 1993, p. 329-350.

Lyness, W.H., Friedle, N.M. and Moore, K.E., Destruction of dopaminergic nerve terminals in nucleus accumbens: effect on d-amphetamine self-administration, *Pharmacol. Biochem. Behav.*, 11, 553, 1979.

Maldonado, R., Dauge, V., Feger, J. and Roques, B.P., Chronic blockade of D2 but not D1 dopamine receptors facilitates behavioral responses to endogenous enkephalins, protected by kelatorphan, administered in the accumbens in rats, *Neuropharmacol.*, 29, 215, 1990.

Mansour, A., Khachaturian, H., Lewis, M.E., Akil, H. and Watson, S.J., Autoradiographic differentiation of mu, delta and kappa opioid receptors in the rat forebrain and midbrain, *J. Neurosci.*, 7, 2445, 1987.

Mansour, A., Meador-Woodruff, J.H., Bunzow, J.R., Civelli, O., Akil, H. and Watson, S.J., Localization of dopamine D2 receptor mRNA and D1 and D2 receptor binding in the rat brain and pituitary: an in situ hybridization-receptor autoradiographic analysis, *J. Neurosci.*, 10, 2587, 1990.

Mansour, A., Meador-Woodruff, J.H., Zhou, Q.-Y., Civelli, O., Akil, H. and Watson, S.J., Jr., A comparison of D1 receptor binding and mRNA in rat brain using receptor autoradiographic and in situ hybridization techniques, *Neuroscience*, 45, 359, 1991.

Mansour, A. and Watson, S.J., Anatomical distribution of opioid receptors in mammalians: an overview, *Handbook of Exp. Pharmacol.*, 104, 79, 1993.

Matthews, R.T. and German, D.C., Electrophysiological evidence for excitation of rat ventral tegmental area dopamine neurons by morphine, *Neuroscience*, 11, 617, 1984.

Meador-Woodruff, J.H., Mansour, A., Grandy, D.K., Damask, S.P., Civelli, O. and Watson, S.J., Jr., Distribution of D5 dopamine receptor mRNA in rat brain, *Neurosci. Lett.*, 145, 209, 1992.

Mello, N.K., Mendelson, J.H., Bree, M.P. and Lucas, S.E., Buprenorphine suppresses cocaine self-administration by rhesus monkeys, *Science*, 245, 859, 1989.

Miles, K., Anthony, D.T., Rubin, L.L., Greengard, P. and Huganir, R.L., Regulation of nicotinic acetylcholine receptor phosphorylation in rat myotubes by forskolin and cAMP, *Proc. Natl. Acad. Sci. USA*, 84, 6591, 1987.

Mitchell, J.B. and Stewart, J., Facilitation of sexual behaviors in the male rat associated with intra-VTA injections of opiates, *Pharmacol. Biochem. Behav.*, 35, 643, 1990.

Mogenson, G.J., Limbic-motor integration, in *Progress in Psychobiology and Physiological Psychology*, Epstein, A.N. and Morrison, A.R., Eds. Academic Press, New York, 1987, p. 117-170.

Mogenson, G.J. and Wu, M., Differential effects on locomotor activity of injections of procaine into mediodorsal thalamus and pedunculopontine nucleus, *Brain Res. Bull.*, 20, 241, 1988.

Mogenson, G.J., Brudzynski, S.M., Wu, M., Yang, C.R. and Yim, C.C.Y., From motivation to action: A review of dopaminergic regulation of limbic-nucleus accumbens-ventral pallidum-pedunculopontine nucleus circuitries involved in limbic-motor integration, in *Limbic Motor Circuits and Neuropsychiatry*, Kalivas, P.W. and Barnes, C.D., Eds. CRC Press, Boca Raton, 1993, p. 193-236.

Montminy, M.R., Gonzalez, G.A. and Yamamoto, K.K., Regulation of cAMP-inducible genes by CREB, *Trends in Neurosci.*, 13, 184, 1990.

Moratalla, R., Robertson, H.A. and Graybiel, A.M., Dynamic regulation of NGFI-A (zif268, egr1) gene expression in the striatum, *J. Neurosci.*, 12, 2609, 1992.

Morgan, J.I. and Curran, T., Stimulus-transcription coupling in the nervous system: Involvement of the inducible protooncogenes fos and jun, *Annu. Rev. Neurosci.*, 14, 421, 1991.

Morris, B.J., Hollt, V. and Herz, A., Dopaminergic regulation of striatal proenkephalin mRNA and prodynorphin mRNA: Contrasting effects of D1 and D2 antagonists, *Neuroscience*, 25, 525, 1988.

Mudge, A.W., Leeman, S.E. and Fischback, C.D., Enkephalin inhibits release of substance P from sensory neurons in culture and decreases action potential duration, *Proc. Natl. Acad. Sci. USA*, 76, 526, 1979.

Nakayama, H., Okuda, H. and Nakashima, T., Phosphorylation of rat brain nicotinic acetylcholine receptor by cAMP-dependent protein kinase in vitro, *Mol. Brain Res.*, 20, 171, 1993.

Napier, T.C., Dopamine receptors in the ventral pallidum regulate circling induced by opioids injected into the ventral pallidum, *Neuropharmacol.*, 31, 1127, 1992.

Nestler, E.J. and Tallman, J.F., Chronic morphine treatment increases cyclic AMP-dependent protein kinase activity in the rat locus coeruleus, *Mol. Pharmacol.*, 33, 127, 1988.

Nestler, E.J., Terwilliger, R.Z., Walker, J.R., Sevarino, K.A. and Duman, R.S., Chronic cocaine treatment decreases levels of the G protein subunits G_{ia} and G_{oa} in discrete regions of rat brain, *J. Neurochem.*, 55, 1079, 1990.

Nestler, E.J., Molecular mechanisms of drug addiction, *J. Neurosci.*, 12, 2439, 1992.

North, R.A. and Williams, J.T., On the potassium conductance increased by opioids in rat locus coeruleus neurones, *J. Physiol.*, 364, 265, 1985.

North, R.A., Williams, J.T., Suprenant, A. and Christie, M.J., μ and δ receptors belong to a family of receptors that are coupled to potassium channels, *Proc. Natl. Acad. Sci. USA*, 84, 5487, 1987.

North, R.A., Opioid receptors and ion channels, in *Neurobiology of Opioids*, Almeida, O.F.X. and Shippenberg, T.S., Eds. Springer-Verlag, New York, 1991, p. 141-150.

Olds, M.E., Reinforcing effects of morphine in the nucleus accumbens, *Brain Res.*, 237, 429, 1982.

Persico, A.M., Schindler, C.W., O'Hara, B.F., Brannock, M.T. and Uhl, G.R., Brain transcription factor expression: effects of acute and chronic amphetamine and injection stress, *Mol. Brain Res.*, 20, 91, 1993.

Pert, A. and Sivit, C., Neuroanatomical focus for morphine and enkephalin-induced hypermotility, *Nature (London)*, 265, 645, 1977.

Phillips, A.G., LePaine, F.G. and Fibiger, H.C., Dopaminergic mediation of reward produced by direct injection of enkephalin into the ventral tegmental area of the rat, *Life Sci.*, 33, 2505, 1983.

Pickel, V.M., Towle, A.C., Joh, T.H. and Chan, J., Gamma-aminobutyric acid in the medial nucleus accumbens: Ultrastructural localization in neurons receiving monosynaptic input from catecholaminergic afferents, *J. Comp. Neurol.*, 272, 1, 1988.

Pickel, V.M., Chan, J. and Sesack, S.R., Cellular basis for interactions between catecholaminergic afferents and leu-enkephalin-like immunoreactivity in rat caudate-putamen, *J. Neurosci. Res.*, 31, 212, 1992.

Randrup, A. and Munkvad, I., Brain dopamine and amphetamine induced stereotyped behavior, *Acta Pharmacol. Toxicol.*, 25 (suppl.4), 62, 1967.(Abstract)

Raymond, L.A., Blackstone, C.D. and Huganir, R.L., Phosphorylation of amino acid neurotransmitter receptors in synaptic plasticity, *Trends in Neurosci.*, 16, 147, 1993.

Reid, M., Herrera-Marschitz, M., Hokfelt, T., Terenius, L. and Ungerstedt, U., Differential modulation of striatal dopamine release by intranigral injection of gamma-aminobutyric acid (GABA) dynorphin A and substance P, *Eur. J. Pharmacol.*, 147, 411, 1988.

Reid, M.S., O'Connor, W.T., Herrera-Marschitz, M. and Ungerstedt, U., The effects on intranigral GABA and dynorphin A injections on striatal dopamine and GABA release: evidence that dopamine provides inhibitory regulation of striatal GABA neurons via D_2 receptors, *Brain Research*, 519, 255, 1990.

Reiner, A. and Andersen, K.D., The patterns of neurotransmitter and neuropeptide co-occurrence among striatal projection neurons: conclusions based on recent findings, *Brain Res. Rev.*, 15, 251, 1990.

83

Roberts, D.C.S. and Koob, G.F., Disruption of cocaine self-administration following 6-hydroxydopamine lesions of the ventral tegmental area in rats, *Pharmacol. Biochem. Behav.*, 17, 901, 1982.

Robertson, G.S., Vincent, S.R. and Fibiger, H.C., D_1 and D_2 dopamine receptors differentially regulate c-fos expression in striatonigral and striatopallidal neurons, *Neuroscience*, 49, 285, 1992a.

Robertson, H.A., Paul, M.L., Maratell, R. and Graybiel, A.M., Expression of the immediate early gene c-fos in basal ganglia: induction by dopaminergic drugs, *Can. J. Neurol. Sci.*, 18, 380, 1992b.

Robinson, T.E. and Becker, J.B., Enduring changes in brain and behavior produced by chronic amphetamine administration: A review and evaluation of animal models of amphetamine psychosis, *Brain Res. Rev.*, 11, 157, 1986.

Sabol, S.L., Yoshikawa, K. and Hong, J.S., Regulation of methionine-enkephalin precursor RNA in rat striatum by haloperidol and lithium, *Biochem. Biophys. Res. Commun.*, 113, 391, 1983.

Savasta, M., Dubois, A., Benavides, J. and Scatton, B., Different plasticity changes in D_1 and D_2 receptors in rat striatal subregions following impairment of dopaminergic transmission, *Neurosci. Lett.*, 85, 119, 1988.

Schinelli, S., Paolillo, M. and Corona, G.L., Modulation of dopamine-induced cAMP production in rat striatal cultures by the calcium ionophore A23187 and by phorbol-12-myristate-13-acetate, *Mol. Brain Res.*, 21, 162, 1994.

Schmauss, C., Yassouridis, A. and Emrich, H.M., Antipsychotic effect of buprenorphine in schizophrenia, *Am. J. Psychiat.*, 144, 1340, 1987.

Schoffelmeer, A.N.M., Hansen, H.A., Stoof, J.C. and Mulder, A.H., Inhibition of dopamine-stimulated cAMP efflux from rat neostriatal slices by activation of μ- and δ-opioid receptors: a permissive role for D_2 dopamine receptors, *Eur. J. Pharmacol.*, 118, 363, 1985.

Schoffelmeer, A.N.M., Hogenboom, F. and Mulder, A.H., Inhibition of dopamine-sensitive adenylate cyclase by opioids. Possible involvement of physically associated mu and delta opioid receptors, *Naunyn-Schmiedeberg's Arch Pharmacol*, 335, 278, 1987.

Schoffelmeer, A.N.M., Rice, K.C., Jacobson, A.E., Van Gelderen, J.G., Hogenboom, F., Heijna, M.H. and Mulder, A.H., Mu-, delta-, and kappa-opioid receptor-mediated inhibition of neurotransmitter release and adenylate cyclase activity in rat brain slices: studies with fentanyl isothiocyanate, *Eur. J. Pharmacol.*, 154, 169, 1988.

Schottenfeld, R.S., Pakes, J., Ziedonis, D. and Kosten, T.R., Buprenorphine: dose-related effects on cocaine and opioid use in cocaine-abusing, opioid-dependent humans, *Biol. Psychiat.*, 34, 66, 1993.

Seeman, P. and Van Tol, H.H.M., Dopamine receptor pharmacology, *Curr. Opin. Neurol.*, 6, 602, 1993.

Sesack, S.R. and Pickel, V.M., Dual ultrastructural localization of enkephalin and tyrosine hydroxylase immunoreactivity in rat ventral tegmental area: multiple substrates for opiate-dopamine interactions, *J. Neurosci.*, 12, 1335, 1992.

Sheng, M. and Greenberg, M.E., The regulation and function of c-fos and other immediate early genes in the nervous system, *Neuron*, 4, 477, 1990.

Sheng, M., Thompson, M.A. and Greenberg, M.E., CREB: a Ca^{+2}-regulated transcription factor phosphorylated by calmodulin-dependent kinases, *Science*, 252, 1427, 1991.

Shippenberg, T.S., Conditioning of opioid reinforcement: Neuroanatomical and neurochemical substrates, *Ann. N. Y. Acad. Sci.*, 654, 347, 1992.

Sibley, D.R., Benovic, J.L., Caron, M.G. and Lefkowitz, R.J., Phosphorylation of cell surface receptors: a mechanism for regulating transduction pathways, *Endocr. Rev.*, 9, 38, 1988.

Sibley, D.R. and Monsma, F.J., Jr., Molecular biology of dopamine receptors, *Tr. Pharmacol. Sci.*, 13, 61, 1992.

Sivam, S.P., Strunk, C., Smith, D.R. and Hong, J.S., Preproenkephalin-A gene regulation in the rat striatum: influence of lithium and haloperidol, *Mol. Pharmacol.*, 30, 186, 1986.

Smith, J.A.M., Loughlin, S.E. and Leslie, F.M., κ-opioid inhibition of [^{3}H]dopamine release from rat ventral mesencephalic dissociated cell cultures, *Mol. Pharmacol.*, 42, 575, 1992.

Sokoloff, P., Giros, B., Martres, M.-P., Bouthenet, M.-L. and Schwartz, J.-C., Molecular cloning and characterization of a novel dopamine receptor (D_3) as a target for neuroleptics, *Nature*, 347, 146, 1990.

Sokoloff, P., Martres, M.-P., Giros, B., Bouthenet, M.-L. and Schwartz, J.-C., The third dopamine receptor (D3) as a novel target for antipsychotics, *Biochem. Pharmacol.*, 43, 659, 1992.

Sorg, B.A., Chen, S.-Y. and Kalivas, P.W., Time course of tyrosine hydroxylase expression following behavioral sensitization to cocaine, *J. Pharmacol. Exp. Ther.*, 266, 424, 1993.

Starr, M.S., Multiple opiate receptors may be involved in suppressing γ-aminobutyrate release in substantia nigra, *Life Sci.*, 16, 2249, 1985.

Stinus, L., Winnock, M. and Kelley, A.E., Chronic neuroleptic treatment and mesolimbic dopamine denervation induce behavioral supersensitivity to opiates, *Psychopharmacol.*, 85, 323, 1985.

Stinus, L., Nadaud, D., Deminiere, J.-M., Jauregui, J., Hand, T.T. and le Moal, M., Chronic flupentixol treatment potentiates the reinforcing properties of systemic heroin administration, *Biol. Psychiat.*, 26, 363, 1989.

Striplin, C. and Kalivas, P.W., Robustness of G protein changes in cocaine sensitization shown with immunoblotting, *Synapse*, 14, 10, 1993.

Striplin, C.D. and Kalivas, P.W., Correlation between behavioral sensitization to cocaine and G protein ADP-ribosylation in the ventral tegmental area, *Brain Research*, 579, 181, 1992.

Sunahara, R.K., Niznik, H.B., Weiner, D.M., Stormann, T.M., Brann, M.R., Kennedy, J.L., Gelernter, J.E., Rozmahel, R., Yang, Y., Israel, Y. and Seeman, P., Human dopamine D1 receptor encoded by an intronless gene on chromosome 5, *Nature*, 347, 80, 1990.

Sunahara, R.K., Guan, H.-C., O'Dowd, B.F., Seeman, P., Laurier, L.G., Ng, G., George, S.R., Torchia, J., Van Tol, H.H.M. and Niznik, H.B., Cloning of the gene for a human dopamine D_5 receptor with higher affinity for dopamine than D_1, *Nature*, 350, 614, 1991.

Surmeier, D.J., Reiner, A., Levine, M.S. and Ariano, M.A., Are neostriatal dopamine receptors co-localized? *Trends in Neurosci.*, 16, 229, 1993.

Swanson, L.W., The projections of the ventral tegmental area and adjacent regions: A combined fluorescent retrograde tracer and immunofluorescence study in the rat, *Brain Res. Bull.*, 9, 321, 1982.

Swerdlow, N.R. and Koob, G.F., Lesions of the dorsomedial nucleus of the thalamus, medial prefrontal cortex and pedunculopontine nucleus: Effects on locomotor activity mediated by nucleus accumbens-ventral pallidal circuitry, *Brain Res.*, 412, 233, 1987.

Tan, D.-P. and Tsou, K., Intranigral injection of dynorphin in combination with substance P on striatal dopamine metabolism in the rat, *Brain Res.*, 443, 310, 1988.

Thureson-Klein, A.K. and Klein,.R.L., Exocytosis from neuronal large dense-cored vesicles, *Intern. Rev. Cytol.*, 121, 67, 1990.

Tsunoo, A.M., Yoshii, M. and Narahashi, T., Block of calcium channels by enkephalin and somatostatin in neuroblastoma-glioma hybrid NGT 108-15 cells, *Proc. Natl. Acad. Sci. USA*, 83, 9832, 1986.

Uchimura, N., Higashi, H. and Nishi, S., Hyperpolarizing and depolarizing actions of dopamine via D-1 and D-2 receptors on nucleus accumbens neurons, *Brain Res.*, 375, 368, 1986.

Vaccarino, F.J., Bloom, F.E. and Koob, G.F., Blockade of nucleus accumbens opiate receptors attenuates intravenous heroin reward in the rat, *Psychopharmacol.*, 85, 37, 1985.

Van Tol, H.H.M., Bunzow, J.R., Guan, H.-C., Sunahara, R.K., Seeman, P., Niznik, H.B. and Civelli, O., Cloning of the gene for a human dopamine D_4 receptor with high affinity for the antipsychotic clozapine, *Nature*, 350, 610, 1991.

Vernier, P., Julien, J.-F., Rataboul, P., Fourrier, O., Feuerstein, C. and Mallet, J., Similar time course changes in striatal levels of glutamic acid decarboxylase and proenkephalin mRNA following dopaminergic deafferentation in the rat, *J. Neurochem.*, 51, 1375, 1988.

Vezina, P. and Stewart, J., Conditioning and place-specific sensitization of increases in activity induced by morphine in the VTA, *Pharmacol. Biochem. Behav.*, 20, 925, 1984.

Vezina, P. and Stewart, J., Amphetamine administered to the ventral tegmental area but not to the nucleus accumbens sensitizes rats to systemic morphine: lack of conditioned effects, *Brain Research*, 516, 99, 1990.

Volavka, J., Bauman, J., Pevnick, J., Reker, D., James, B. and Cho, D., Short-term hormonal effects of naloxone in man, *Psychoneuroendo.*, 5, 225, 1980.

Voorn, P., Roest, G. and Groenewegen, H.J., Increase of enkephalin and decrease of substance P immunoreactivity in the dorsal and ventral striatum of the rat after midbrain 6-hydroxydopamine lesions, *Brain Res.*, 412, 391, 1987.

Walker, J.M., Thompson, L.A., Frascella, J. and Friederich, M.W., Opposite effects of μ and κ opiates on the firing-rate of dopamine cells in the substantia nigra of the rat, *Eur. J. Pharmacol.*, 134, 53, 1987.

Werner, T.E., Smith, S.G. and Davis, W.M., A dose-response comparison between methadone and morphine self-administration, *Psychopharmacol.*, 47, 209, 1976.

Williams, J.T., Egan, T.M. and North, R.A., Enkephalin opens potassium channels on mammalian central neurones, *Nature*, 299, 74, 1982.

Williams, N. and Clouet, D.H., The effect of acute opioid administration on the phosphorylation of rat striatal synaptic membrane proteins, *J. Pharmacol. Exp. Ther.*, 220, 278, 1982.

Wise, R.A. and Rompre, P.P., Brain dopamine and reward, *Annu. Rev. Psychol.*, 40, 191, 1989.

Wood, P.L., Stotland, M., Richard, J.W. and Rackham, A., Actions of mu, kappa, delta and agonist/antagonist opiates on striatal dopaminergic function, *J. Pharmacol. Exp. Ther.*, 215, 697, 1980.

Zaborszky, L., Alheid, G.F. and Heimer, L., Mapping of transmitter-specific connections: simultaneous demonstration of anterograde degeneration and changes in the immunostaining pattern induced by lesions, *J. Neurosci. Meth.*, 14, 255, 1985.

Zahm, D.S., Zaborszky, L., Alones, V.E. and Heimer, L., Evidence for the coexistence of glutamate decarboxylase and met-enkephalin immunoreactivities in axon terminals of rat ventral pallidum, *Brain Res.*, 325, 317, 1985.

Zahm, D.S. and Brog, J.S., On the significance of subterritories in the "accumbens" part of the rat ventral striatum, *Neuroscience*, 50, 751, 1992.

Zhou, Q.-Y., Grandy, D.K., Thambi, L., Kushner, J.A., Van Tol, H.H.M., Cone, R., Pribnow, D., Salon, J., Bunzow, J.R. and Civelli, O., Cloning and expression of human and rat D_1 dopamine receptors, *Nature*, 347, 76, 1990.

CHAPTER 3

SEROTONERGIC MODULATION OF MIDBRAIN
DOPAMINE SYSTEMS

Mark D. Kelland and Louis A. Chiodo

I. INTRODUCTION

The presence of a major ascending projection from serotonin (5-HT)-containing neurons of the rostral raphe nuclei to dopamine (DA)-rich areas of the brain has been known for over a decade. [Dray et al., 1976; Parent, et al., 1981; Van der Kooy and Hattori, 1980; for review see Carpenter and Sutin, 1983; Tork, 1985] 5-HT projections innervate not only midbrain regions containing DA cell bodies, but also DA terminal regions, such as the neostriatum and areas of prefrontal cortex (PFC). Although Kalivas [1993] recently provided a concise review of 5-HT modulation of DA function, most reviews of midbrain DA systems largely ignore the influence of 5-HT, and recent texts on the neuropharmacology and psychoactive properties of 5-HT and of drugs which interact with 5-HT systems have said little about the regulation of DA. [Rech and Gudelsky, 1988; Whitaker-Azmitia and Peroutka, 1990]

More recently, however, the influence of 5-HT on midbrain DA systems has received significant attention. While many of these studies have presented a complicated picture in which 5-HT exerts a modulatory influence on DA function, one promising line of research has significant clinical relevance: the observation that atypical antipsychotics often share both DA and 5-HT antagonist properties. [for review see Deutch et al., 1991; Glennon, 1990; Greenshaw, 1993; Meltzer, 1989, 1992; Tricklebank, 1989] This has led to the proposal of a DA/5-HT hypothesis of schizophrenia. [Meltzer, 1989] In this chapter we will attempt to provide a general overview of the influence exerted by 5-HT on the function of midbrain DA systems as demonstrated by a variety of experimental techniques.

II. ANATOMY

A. 5-HT CONTAINING PATHWAYS

The anatomy of 5-HT pathways has been reviewed previously. [Carpenter and Sutin, 1983; Tork, 1985, 1990] 5-HT-containing neurons of the rostral raphe nuclei send a major ascending projection throughout the basal ganglia and cerebral cortex. Among those sites receiving a 5-HT input are the substantia nigra and ventral tegmental area (VTA) (DA-cell rich areas A9 and A10 of Dahlstrom and Fuxe [1964]), as well as the DA-terminal regions of the neostriatum and PFC. [Dray et al., 1976; Herve et al., 1987; Parent et al., 1981; Soghomonian et al., 1987; Van der Kooy and Hattori, 1980] It has been further demonstrated that the 5-HT projection to the substantia nigra arises primarily from the dorsal raphe nucleus. [Fibiger and Miller, 1977; Wirtshafter et al., 1987] These extensive connections from 5-HT-containing neurons to both DA-cell body regions and their terminal regions obviously provide the 5-HT system with the potential for significantly influencing the activity of the midbrain DA systems.

B. 5-HT RECEPTOR SUBTYPES IN DA-RICH AREAS

5-HT receptors have been divided into a number of subtypes, and anatomical studies have begun to identify the distribution of these receptor subtypes within the basal ganglia and PFC. [Palacios et al., 1990; Peroutka, 1988] 5-HT neurons within the dorsal raphe appear to have somatodendritic autoreceptors of the 5-HT$_{1A}$ type [Blier and De Montigny, 1987; Blier et al., 1987; Cox et al., 1993; De Montigny and Aghajanian, 1977; Marcinkiewicz et al., 1984; Pazos and Palacios, 1985; Sprouse and Aghajanian, 1987], whereas the A9 and A10 regions contain high percentages of the 5-HT$_{1B}$ receptor subtype. [Marcinkiewicz et al., 1984; Pazos and Palacios, 1985] The relative lack of other 5-HT receptor subtypes suggests that the majority of postsynaptic 5-HT receptors in the substantia nigra and VTA are of the 5-HT$_{1B}$ subtype. [Fischette et al., 1987; Marcinkiewicz et al., 1984; Pazos and Palacios, 1985] Regarding the 5-HT$_1$ receptor subtype, DA-terminal regions also contain high concentrations of the 5-HT$_{1B}$ receptor. [Fischette et al., 1987; Marcinkiewicz et al., 1984; Pazos and Palacios, 1985] However, the neocortex also has an enriched density of the 5-HT$_{2A}$ receptor subtype, possibly suggesting a special role for 5-HT$_{2A}$ receptor modulation of mesocortical DA systems. [Fischette et al., 1987; Palacios et al., 1990] A recent study by Grossman et al. [1993] has also demonstrated significant concentrations of 5-HT$_4$ receptors in various basal ganglia nuclei.

C. DEVELOPMENTAL COMPENSATION BY 5-HT NEURONS FOLLOWING NEONATAL DA LESIONS

That 5-HT and DA are intimately related within the neostriatum is demonstrated by an interesting line of developmental research. Stachowiak et al. [1984] first demonstrated that neonatal lesions of the DA system, by intraventricular administration of the selective DA neurotoxin 6-hydroxydopamine (6-OHDA), result in sprouting and proliferation of 5-HT projections to the neostriatum, particularly in subregions which are typically DA-rich but 5-HT-poor. Similarly, in the weaver mutant mouse, which contains a recessive mutation resulting in the loss of DA neurons, there is a compensatory increase in 5-HT, with the greatest increases in 5-HT content occurring in the areas of greatest DA-depletion. [Stotz et al., 1993] Further studies have confirmed the specific role of DA-depletion in this phenomenon [Luthman et al., 1987], have demonstrated that the 5-HT hyperinnervation arises primarily from dorsal raphe afferents [Berger et al., 1985] and have also shown increases in neostriatal levels of both 5-HT_{1B} and 5-HT_{2A} receptors. [Radja et al., 1993] Based on *in vivo* neurochemical studies, neonatal 6-OHDA lesions also appear to increase the capacity for 5-HT release in the neostriatum. [Jackson and Abercrombie, 1992]

Since neonatal 6-OHDA lesions do not induce the severe behavioral disorders seen in adult rats with comparable lesions, it was initially proposed that 5-HT hyperinnervation compensated for the decreased DA levels. [Bruno et al., 1984] However, Bruno et al. [1987] demonstrated this was not the case, since DA lesions at a later age resulted in behavioral sparing without 5-HT hyperinnervation and combined neonatal DA and 5-HT lesions also resulted in behavioral sparing without 5-HT hyperinnervation. Thus, the significance of this interesting phenomenon remains unclear. Further clouding the issue is the recent observation that 6-OHDA lesions in adult rats also results in significant sprouting of 5-HT fibers in the neostriatum, and a concomitant increase in levels of 5-HT and the 5-HT metabolite 5-hydroxyindoleacetic acid. [Zhou et al., 1991] The latter study is in contrast to the general conclusion of the earlier studies that 5-HT hyperinnervation is strictly a developmental phenomenon dependent on neonatal DA lesions.

III. ELECTROPHYSIOLOGICAL STUDIES

A. 5-HT MODULATION OF THE BASAL ACTIVITY OF DA NEURONS

Since it has been difficult to identify the direct effects of 5-HT on midbrain DA neurons, the functional significance of ascending 5-HT projections to these cells has remained unclear. Either iontophoretic administration of 5-HT or electrical stimulation of 5-HT pathways has reliable inhibitory effects on midbrain DA neurons and their terminal regions. [Aghajanian and Bunney, 1974; Crossman et al., 1974; Davies and Tongroach, 1978; Dray et al., 1976; Fibiger and Miller, 1977; Kostowski et al., 1969] More recently, we have observed that electrical stimulation of the dorsal raphe inhibits only a subset of nigrostriatal DA neurons, referred to as slow DA neurons (firing rates < 4 spikes/sec), whereas all mesoaccumbens DA were inhibited by raphe stimulation. [Kelland et al., 1990a, 1993] Our initial observation that stimulation of 5-HT inputs preferentially inhibited slow DA neurons led us to speculate that the low firing rates of these cells might be due to the inhibitory 5-HT projection. [Kelland et al., 1990a] However, the more appropriate conclusion appears to be that slow DA neurons are more likely to be under a variety of afferent regulation, since excitatory inputs from the pedunculopontine tegmental nucleus preferentially affect slow DA neurons in both the nigrostriatal and mesoaccumbens DA systems. [Kelland et al., 1993] Nonetheless, these data may help to explain why previous studies on the responsiveness of DA neurons to 5-HT (studies performed primarily on nigrostriatal DA neurons) have met with limited success. According to our results, only approximately half of all spontaneously active nigrostriatal DA neurons are influenced by 5-HT projections. [Kelland et al., 1990a] Of the remaining half there will, of course, be some variability in responsiveness. Thus, only a small subset of DA neurons are likely to demonstrate a robust response to increased 5-HT neurotransmission.

As with electrical stimulation of 5-HT projections or iontophoretic administration of 5-HT directly onto DA neurons, studies utilizing 5-HT agonists have demonstrated equivocal results. The 5-HT$_{1A}$ agonist 8-hydroxy-2-(di-n-propylamino)tetralin (DPAT) has been reported to either excite or fail to influence the firing rate of DA neurons at low doses, while causing inhibition at very high doses. [Lum and Piercey, 1988; Sinton and Fallon, 1988] Other 5-HT$_{1A}$ agonists also either excited or had no effect on DA neurons. [Lum and Piercey, 1988; Sinton and Fallon, 1988] In contrast, 5-HT$_{1B}$ agonists exert a reliable, though weak, inhibitory effect on the firing rate of midbrain DA neurons. [Kelland, et al., 1990a; Sinton and Fallon, 1988] When DA neurons are divided into slow and fast cells (as described above), 5-HT$_{1A}$ agonists reliably induce an increase in the firing rate of slow DA neurons. [Kelland et al., 1990a] This effect appears to involve disinhibition as a result of 5-HT$_{1A}$ agonist-induced inhibition of 5-HT neurons, since prior depletion of 5-HT by the neurotoxin 5,7-dihydroxytryptamine (5,7-DHT) eliminated the response. [Kelland et al., 1990a] Recent studies by Arborelius et al. [1993a,b] have shown that DPAT may exert a greater influence on A10 DA neurons, as opposed to A9 DA cells, in regard both to causing an increase in burst-firing and to increasing DA release in PFC.

The influence of 5-HT on midbrain DA neurons appears to be phasic, rather than tonic. Depletion of 5-HT by either p-chlorophenylalanine (PCPA) or 5,7-DHT had little effect on the basal activity of DA neurons. [Kelland et al., 1990a,b] Similarly, disruption of 5-HT transmission by 3,4-methylenedioxymethamphetamine (MDMA) had no significant effects on DA cell basal activity. [Kelland et al., 1989d; for detailed discussion of MDMA see below] In a biochemical study, Fibiger and Miller [1977] also found that PCPA-induced depletion of 5-HT did not alter the subsequent ability of alpha-methyl-p-tyrosine to reduce DA levels in the neostriatum. [see, however, Herve et al., 1979] Thus, an alternative explanation for the difficulty in identifying reliable effects of 5-HT or 5-HT-active compounds may be the lack of any tonic modulation which might be altered by 5-HT antagonist administration. When combined with observations that the primary role of 5-HT might be presynaptic regulation of other neurotransmitter systems (e.g., glutamate and GABA [Aghajanian and Bunney, 1974; Johnson et al., 1992]), it is understandable that the influence of 5-HT on DA cell basal activity has been difficult to elucidate.

B. 5-HT MODULATION OF THE PHARMACOLOGICAL AND PHYSIOLOGICAL RESPONSIVENESS OF DA NEURONS

An important component of the regulation of midbrain DA neuronal responsiveness appears to be dendritic release of DA onto somatodendritic autoreceptors on the DA cells. [for review see Chiodo, 1988, 1992; Grace and Bunney, 1985] Several studies suggest that 5-HT inputs from the dorsal raphe influence dendritic release of DA and, thus, autoreceptor regulation of DA cell responsiveness. DA neurons have a dendritic calcium conductance typical of cells exhibiting dendritic release of neurotransmitter [Chiodo and Kapatos, 1987, 1992; Grace and Onn, 1989; Llinas et al., 1984], and 5-HT facilitates a dendritic calcium conductance in DA neurons. [Nedergaard et al., 1988] In addition, dorsal raphe stimulation reduces the somatodendritic component of neostriatal-evoked antidromic responses in substantia nigra DA neurons, demonstrating a decrease in somatodendritic excitability. [Trent and Tepper, 1991] That the latter results are directly dependent upon 5-HT release caused by raphe stimulation is shown by the observations that either the 5-HT antagonist metergoline or depletion of 5-HT by PCPA could eliminate the effect of raphe stimulation on DA cell somatodendritic excitability. [Trent and Tepper, 1991] Finally, 5-HT also appears to modulate GABAergic terminals which hyperpolarize DA neurons. Johnson et al. [1992] demonstrated that 5-HT reduced the amplitude of $GABA_B$ synaptic potentials in DA neurons, and that this effect was mediated by $5\text{-}HT_{1B}$ receptors (but not $5\text{-}HT_{1A}$ or $5\text{-}HT_{2A}$).

Given the effects of 5-HT on DA cell membranes, changes in the responsiveness of DA cells would be expected. However, only limited data are available on this topic. The ability of DA receptor agonists to inhibit the firing rate of DA cells is typically dependent on the basal firing rate of the cell being examined. [Bunney et al., 1973; Clark et al., 1985; Kelland et al., 1988; White and Wang, 1984] Depletion of 5-HT eliminates the rate-dependent inhibition of DA neurons by the D_2/D_3 agonist quinpirole, while having no significant effect on the basal activity of these cells. [Kelland et al., 1990a] Interestingly, the D_1 agonist SKF 38393 also disrupts the rate-dependent inhibition of DA neurons by quinpirole [Kelland et al., 1988], and SKF 38393 has been shown to inhibit 5-HT release. [Benkirane et al., 1987] In contrast, however, 5-HT depletion had no effect on the inhibition of DA neurons by the highly potent D_2 agonist (+)-4-propyl-9-hydroxynaphthoxazine [Kelland et al., 1990b]. 5-HT also modulates the responsiveness of DA neurons to glutamate, significantly reducing glutamate-induced excitation even in the absence of measurable effects by 5-HT alone. [Aghajanian and Bunney, 1974] Whether the latter observation involves 5-HT effects on the DA cell membrane, or whether it is an effect on glutamate terminals, is not known at the present time. However, 5-HT does modulate the responsiveness of DA neurons to GABA by a presynaptic mechanism. [Johnson et al., 1992]

C. IS SENSORY INPUT TO DA NEURONS MEDIATED BY 5-HT?

The activity of midbrain DA neurons is altered by a variety of peripheral sensory stimuli, and destruction of forebrain feedback pathways does not eliminate these responses. [Chiodo et al., 1979, 1980; Feger et al., 1978; Grace et al., 1980; Hommer and Bunney, 1980; Kelland et al., 1989b, 1991, 1993; Mantz et al., 1989; Schultz and Romo, 1987; Strecker and Jacobs, 1987] Since 5-HT neurons respond to similar sensory stimuli, and then project to DA cells, it is possible that 5-HT pathways mediate the responsiveness of DA neurons to peripheral sensory stimulation. [Aghajanian et al., 1978; Heym et al., 1982; Jacobs et al., 1990; Rasmussen et al., 1986]

In rats, electrical stimulation of the sciatic nerve inhibits the activity of nigrostriatal DA neurons. [Hommer and Bunney, 1980; Kelland et al., 1989b, 1991, 1993] Although we have reported that sciatic nerve stimulation does not inhibit mesoaccumbens DA neurons [Kelland et al., 1993], a more recent experiment has determined that sciatic nerve stimulation also inhibits mesoaccumbens DA neurons, but lower stimulation currents must be utilized. [Chiodo and Kelland, unpublished observations] As described above, 5-HT projections to DA neurons are typically inhibitory and, therefore, could mediate the inhibition caused by sciatic nerve stimulation. Thus, we have used this paradigm to evaluate the role of 5-HT in sensory modulation of DA neurons.

Inhibitory 5-HT projections from the dorsal raphe are clearly involved in the effects of sciatic nerve stimulation on DA neurons. [Kelland et al., 1993] However, the role of 5-HT has been difficult to elucidate because different methods of depleting endogenous 5-HT have variable effects. Following depletion of brain 5-HT with either 5,7-DHT or PCPA, significantly fewer DA cells were inhibited by sciatic nerve stimulation. Further, the D_1 agonist SKF 38393 was no longer able to enhance this inhibition in 5-HT-depleted animals. The ability of SKF 38393 to enhance sciatic nerve stimulation-induced inhibition of DA neurons may be related to a D_1-mediated reduction of 5-HT release. [Benkirane et al., 1987] Supporting this contention is the observation that DPAT, at doses which completely suppress the firing of 5-HT neurons in the dorsal raphe [Blier and De Montigny, 1987; Blier et al., 1987; Cox et al., 1993; Lum and Piercey, 1988; Sinton and Fallon, 1988], also enhances the inhibition of DA neurons. [Kelland et al., 1993] However, these data are difficult to interpret, because the acute reduction of 5-HT (i.e., DPAT-induced suppression of firing rate or the proposed SKF 38393-induced blockade of release) has the opposite effect of chronic reduction of 5-HT (i.e., 5,7-DHT- or PCPA-induced depletion). Negative results with trifluoromethylphenylpiperazine, which blocks the primary postsynaptic 5-HT receptors in the substantia nigra (5-HT$_{1B}$ receptors [Marcinkiewicz et al., 1984; Pazos and Palacios, 1985]), further cloud the issue, because the reduction of 5-HT release would result in decreased postsynaptic 5-HT receptor activation. Thus, trifluoromethylphenylpiperazine and DPAT were expected to have similar effects, but they did not. Although we cannot yet explain these differences, possible underlying factors contributing to differences between acute vs. chronic treatments include compensatory changes in the central nervous system following chronic 5-HT depletion and/or the difference between impulse-dependent changes in 5-HT availability vs. actual depletion of this neurotransmitter.

D. THE INFLUENCE OF 5-HT$_{2A}$ AND 5-HT$_3$ RECEPTORS: THE MESOCORTICAL DA SYSTEM AND SCHIZOPHRENIA

The medial prefrontal cortex (mPFC) is an import DA projection site involved in cognition and emotion and which is implicated in psychotic disorders such as schizophrenia. [see Andreasen, 1990; Carpenter and Sutin, 1983; Lynch, 1992; Weinberger et al., 1988] The development of schizophrenia has recently been hypothesized to involve a chronic deficiency in striatal glutamate transmission due to decreased activity in regions of PFC which project to the dorsal striatum (the caudate/putamen nuclei) and/or the ventral striatum (the nucleus accumbens). [for theoretical review see Carlsson, 1988; Grace, 1991] Reduced PFC activity which might underlie this proposed deficit in glutamatergic function in the striatum of schizophrenic patients has been reviewed [Andreasen, 1990; Lynch, 1992], and prefrontal structural and functional deficits have also been reported in patients with schizotypal personality disorders. [Raine et al., 1992] It is important to note that this new theory does not contradict the DA theory of schizophrenia, but rather raises important questions about the primary dysfunction causing this disorder. Not only is the striatum a major site of DA projections, but so is the PFC. [for review see Bunney and Sesack, 1987; Lynch, 1992; Thierry et al., 1988] Thus, 5-HT modulation of PFC systems also has important implications for understanding the etiology and putative treatments of schizophrenia.

1. mPFC Modulation of DA Cell Activity

In vivo, DA neurons often fire in a bursting pattern. [Chiodo et al., 1984; Clark and Chiodo, 1988; Grace and Bunney, 1984; Kelland et al., 1989a,c] The importance of the burst-firing pattern was first demonstrated by Gonon and Buda [1985]. Utilizing *in vivo* voltammetry, they demonstrated that DA neurons release significantly more DA into the striatum when exposed to stimulation parameters which mimic burst firing, as compared to being stimulated in a more regular pattern (overall pulses/sec were held constant). We have demonstrated that electrical stimulation of the medial forebrain bundle, which contains the majority of ascending DA axons, in a burst-firing pattern increases both the DOPAC/DA ratio in the olfactory tubercle and preproenkephalin mRNA in the striatum. [Bannon et al., 1989]

mPFC activity appears necessary for the maintenance of burst-firing in midbrain DA neurons. Electrical stimulation of mPFC induces burst-firing in DA cells [Gariano and Groves, 1988], whereas local cooling of mPFC significantly reduces bursting in DA neurons. [Svensson and Tung, 1989] That this is a modulatory effect is suggested by the observation that the reduction in burst-firing during mPFC inactivation is not accompanied by any significant effect on the mean firing rate of the DA neurons. [Svensson and Tung, 1989] Since the 5-HT$_{2A}$ receptor antagonist ritanserin has been shown to increase the activity of midbrain DA neurons by blocking 5-HT-induced inhibition of these cells [Ugedo et al., 1989], Svensson et al. [1989] examined whether ritanserin could reverse the effect of mPFC cooling on DA neurons. As expected, ritanserin did block the reduction in burst-firing caused by mPFC inactivation. [Svensson et al., 1989] Thus, cortical modulation of DA cell activity is under the influence of 5-HT afferents. Since cooling of the mPFC would eliminate the possibility of 5-HT interactions at that site, it is likely that ritanserin acted at the cortical terminal regions within the basal ganglia, where 5-HT has been shown to have antagonist modulatory properties on glutamate projections. [Aghajanian and Bunney, 1974; see also Davies and Tongroach, 1978]

2. 5-HT Modulation of mPFC Neurons

The results described in the preceding section suggested a modulatory role for 5-HT in the terminal regions where mPFC neurons project. We will now examine the effects of 5-HT on mPFC neurons. Agonists for either 5-HT$_{2A}$ or 5-HT$_3$ receptors reliably inhibit the spontaneous activity of mPFC neurons. [Ashby and Wang, 1990; Ashby et al., 1989a,b, 1990a, 1991a,b] These effects appear to be specific and, depending on the experimental paradigm, either dose- or current-dependent. [Ashby et al., 1989a, 1990a, 1991a] At low currents, 5-HT$_{2A}$ agonists also potentiated glutamate-induced excitation of mPFC neurons. [Ashby et al., 1990a] Based on increasing evidence that antipsychotic drugs, especially atypical antipsychotics, share both DA and 5-HT antagonist properties (see below, Section IV.), the effects of antipsychotic medication on the suppressant-action of 5-HT agonists on mPFC neurons were examined. Typical antipsychotic drugs, e.g. haloperidol and chlorpromazine, failed to reduce the effects of either 5-HT$_{2A}$ or 5-HT$_3$ agonist-induced inhibition of mPFC neurons. [Ashby and Wang, 1990; Ashby et al., 1989b, 1991b] Actually, haloperidol appeared to potentiate the effects of each receptor subtype agonist. [Ashby and Wang, 1990; Ashby et al., 1989b] In contrast, the atypical antipsychotic clozapine significantly reduced the inhibitory effects of both 5-HT$_{2A}$ and 5-HT$_3$ agonists on mPFC cells. [Ashby and Wang, 1990; Ashby et al., 1989b, 1991b]

As with 5-HT, mesocortical DA projections to mPFC typically have inhibitory actions on mPFC neurons. [Thierry et al., 1988] Given the likely role of PFC in schizophrenia, and the similarity of 5-HT and DA function there, the observation that atypical antipsychotics block 5-HT pathways in PFC raises important issues regarding the role 5-HT has in the etiology and/or manifestation of schizophrenia. However, in an examination of several putative atypical antipsychotics other than clozapine, Ashby et al. [1991b] were unable to duplicate their results. Whether the putative antipsychotics will fail to prove effective in the clinical management of schizophrenic symptoms, or whether clozapine is unique even among atypical antipsychotics remains to be determined. The next section, however, turns to a different electrophysiological model assessing the putative efficacy and typical or atypical nature of antipsychotic drugs.

3. Effects of 5-HT Antagonists and Atypical Antipsychotics in an Electrophysiological Model of Antipsychotic Efficacy

Population studies on the number of spontaneously active midbrain DA neurons, typically measuring the number of cells-per-track in a given brain region, have provided a useful model for determining the antipsychotic potential as well as the potential for extrapyramidal side effects typical of many antipsychotic compounds. [for review see Chiodo, 1988] Chronic treatment with typical antipsychotics, such as haloperidol, reduce the number of spontaneously active DA neurons in both the substantia nigra and the VTA. [Bunney and Grace, 1978; Chiodo and Bunney, 1983, 1985; White and Wang, 1983] In contrast, chronic treatment with atypical antipsychotics, such as clozapine, reduces the number of spontaneously active DA neurons only in the VTA. [Chiodo and Bunney, 1983, 1985; White and Wang, 1983] Although Chiodo and Bunney [1985] first proposed an anticholinergic mechanism as the difference between clozapine and typical antipsychotics, current research has focused on the antiserotonergic properties of putative atypical antipsychotics.

The available literature, testing a variety of $5-HT_{2A}$ receptor antagonists, agrees that these compounds selectively reduce the number of spontaneously active DA neurons in the VTA, but not in the substantia nigra. [Goldstein et al., 1989; Skarsfeldt, 1992; Skarsfeldt and Perregaard, 1990; Sorensen et al., 1993] This profile is indicative of atypical antipsychotic activity, i.e., antipsychotic efficacy with reduced risk of extrapyramidal side effects. In addition, the D_1 antagonist SCH 23390, which also acts as a $5-HT_{2A}$ antagonist [Bischoff et al., 1986], has been reported to selectively reduce the spontaneous activity of VTA DA neurons following chronic treatment. [Goldstein and Litwin, 1988; see, however, Esposito and Bunney, 1989] When combined with further evidence that $5-HT_{2A}$ antagonists can either reverse or block the effects of amphetamine on DA neurons, a strong case can be made for recognizing $5-HT_{2A}$ antagonists as putative atypical antipsychotics. [Goldstein et al., 1989; Sorensen et al., 1992, 1993]

The effects of $5-HT_3$ antagonists in this model are less clear. There have been reports that chronic administration of $5-HT_3$ antagonists selectively reduce spontaneous DA cell activity in the VTA, suggesting potential for these compounds as atypical antipsychotics. [Ashby et al., 1994; Prisco et al., 1992; Rasmussen et al., 1991] However, there are also reports that chronic $5-HT_3$ antagonist administration is not selective in reducing the spontaneous activity of midbrain DA neurons. [Minabe et al., 1991b; Sorensen et al., 1989]

Although the latter studies are indicative of antipsychotic potential, but with a typical profile, the mechanism of action may still be unique. The decreased spontaneous activity of DA neurons following chronic 5-HT$_3$ antagonist administration is not reversed by acute administration of the DA agonist apomorphine, suggesting that, unlike DA antagonist antipsychotic drugs, chronic administration of 5-HT$_3$ antagonists does not reduce DA cell activity via depolarization inactivation. [Ashby et al., 1994; Minabe et al., 1991b; Rasmussen et al., 1991; see, however, Prisco et al., 1992] There are also two studies questioning whether 5-HT$_3$ antagonists have any antipsychotic potential when tested in electrophysiological models. Ashby et al. [1990b] reported that the 5-HT$_3$ receptor antagonist BRL 43694 had no significant effects on the activity of midbrain DA neurons. Also, the 5-HT$_3$ receptor antagonists LY 277359 and granisetron both potentiated the suppressant actions of apomorphine on the activity of VTA DA neurons. [Minabe et al., 1991a] These studies would not support the contention that 5-HT$_3$ receptor antagonists have antipsychotic potential. Thus, the literature seems evenly split on suggesting that 5-HT$_3$ receptor antagonists have potential as either typical or atypical antipsychotics, or they have no antipsychotic potential. A definitive conclusion in this area cannot presently be drawn, but other reviews are also generally skeptical regarding the antipsychotic potential of 5-HT$_3$ antagonists. [Deutch et al., 1991; Greenshaw, 1993; Tricklebank, 1989]

IV. DA/5-HT HYPOTHESIS OF SCHIZOPHRENIA

The preceding section (Section III.D.) reviewed electrophysiological studies which suggest a significant role for 5-HT in modulating DA systems, particularly regarding the effects of 5-HT$_{2A}$ receptor antagonists in the cells-per-track model used to determine the putative clinical efficacy of antipsychotics and their potential for side effects. Further evidence described below, from biochemical and clinical studies, supports the contention that 5-HT modulation of midbrain DA systems (particularly via 5-HT$_{2A}$ receptors) plays a critical role in the etiology and/or manifestation of schizophrenic symptoms. These studies have led to the proposal of a DA/5-HT hypothesis of schizophrenia. [Meltzer, 1989] As with the glutamate theory of schizophrenia described above [see Carlsson, 1988; Grace, 1991], the DA/5-HT theory is a modification, not a refutation, of the DA hypothesis of schizophrenia. And, as also described above, certain aspects of 5-HT modulation of DA activity may involve actions on PFC glutamate projections [see Aghajanian and Bunney, 1974; Svensson et al., 1989], thus tying all three theories together in a complicated picture of the etiology and/or manifestation of schizophrenia. There have been several comprehensive reviews of the role of 5-HT antagonists in the treatment of schizophrenia, so we will merely provide an overview of this research. [Deutch et al., 1991; Glennon, 1990; Greenshaw, 1993; Meltzer, 1989, 1992; Tricklebank, 1989]

A. BIOCHEMICAL STUDIES

Both typical and atypical antipsychotics bind to both DA and 5-HT receptors. As compared to typical antipsychotics, atypical antipsychotics demonstrate relatively weak binding to D_2 receptors and potent binding to 5-HT_{2A} receptors, with the ratio of 5-HT_{2A}/D_2 binding being a valid measure for the classification of typical and atypical antipsychotic drugs. [Matsubara et al., 1993; Meltzer et al., 1989a,b] Although the data regarding acute vs. chronic administration of typical and atypical antipsychotics on the density of 5-HT_{2A} and D_2 receptors are not entirely consistent, again there was a significantly greater relationship between atypical antipsychotics and 5-HT_{2A} receptors, whereas typical antipsychotics had a greater effect on D_2 receptor densities. [Matsubara and Meltzer, 1989; Wilmot and Szczepanik, 1989]. Clozapine, but not haloperidol, also mimicked the effects of 5-HT by inhibiting K^+-stimulated release of [^3H]5-HT in the nucleus accumbens. [Drescher and Hetey, 1988] Thus, an interaction with 5-HT receptors, particularly 5-HT_{2A} receptors, appears to be a characteristic of atypical antipsychotic drugs.

Saller et al. [1990] have demonstrated that 5-HT_{2A} antagonists are able to reverse some of the compensatory responses of the brain to chronic D_2 receptor blockade. This suggests that atypical antipsychotics, by virtue of their 5-HT_{2A} receptor antagonist properties, may actually enhance their own atypical profile by reversing compensatory mechanisms of D_2 receptor blockade which might be involved in the development of extrapyramidal side effects. Clozapine also appears to preferentially affect DA transmission in mesolimbic pathways as opposed to the nigrostriatal pathway. Both clozapine and the 5-HT_3 receptor antagonist DAU 6215 selectively antagonized (+)SKF 10,047-induced DA release in the nucleus accumbens. [Volonte et al., 1992] Also, in 5-HT-denervated rats, clozapine preferentially enhanced DA levels in the nucleus accumbens as compared to the neostriatum. [Chen et al., 1992b] These data demonstrate additional mechanisms by which atypical antipsychotics and/or 5-HT_{2A} antagonists appear useful in the treatment of schizophrenia, with a reduced risk of extrapyramidal side effects.

B. CLINICAL STUDIES

1. 5-HT_{2A} Antagonists as Putative Antipsychotics

Many antipsychotic drugs have an affinity for 5-HT receptors as well. In this section, however, we will concentrate on studies involving compounds relatively selective in their blockade of 5-HT_{2A} receptors, as well as evidence for altered 5-HT function in schizophrenic patients.

Promising clinical results in the treatment of schizophrenia have been obtained with several 5-HT_{2A} antagonists, administered either alone or as an adjunct to typical neuroleptics. Ritanserin, a highly selective 5-HT_{2A} receptor antagonist with little affinity for DA receptors [Awouters et al., 1988], proved to be a useful adjunct in treating schizophrenic symptoms. [Gelders, 1989] The selective 5-HT_{2A} antagonists amperozide, melperone and setoperone were all effective in treating schizophrenia when administered alone. [Axelsson et al., 1991; Bjerkenstedt, 1989; Ceulemans et al., 1985; Christensson and Bjork, 1990] These data clearly demonstrate that 5-HT$_{2A}$ receptor antagonists deserve further research in regards to their role as antipsychotic medication.

The mechanism underlying the effects of 5-HT$_{2A}$ antagonists remains to be determined. However, there has been a postmortem report of decreased 5-HT$_{2A}$ receptors in the PFC of chronic schizophrenics. [Mita et al., 1986] In addition, neuroleptics have been shown to bind to 5-HT$_{2A}$ receptors in human frontal cortex. [Wander et al., 1987] Thus, the PFC is indicated as a possible site of action for 5-HT$_{2A}$ receptor antagonists in the treatment of schizophrenia. In support of this proposal, the selective 5-HT$_{2A}$ antagonist and putative atypical antipsychotic amperozide preferentially increases DA efflux in mPFC as opposed to the neostriatum. [Pehek et al., 1993] Although the majority of the literature supports a role for 5-HT$_{2A}$ receptors in schizophrenia, there are also reports of neuroleptics binding to 5-HT$_{1A}$ receptors in human frontal cortex [Wander et al., 1987] and increased densities of 5-HT$_{1A}$ receptors in PFC of schizophrenic patients. [Hashimoto et al., 1991] The density of 5-HT$_3$ receptors, however, appears to be normal in postmortem brain tissue from schizophrenic subjects. [Abi-Dargham et al., 1993] This latter study is in general agreement with the lack of promise shown to date in tests of the antipsychotic activity of 5-HT$_3$ antagonists. [Deutch et al., 1991; Greenshaw, 1993; Tricklebank, 1989]

2. Atypical Antipsychotics: Is 5-HT$_{2A}$ or D$_1$ Antagonism Responsible for Their Unique Profile?

The prototypic atypical antipsychotic medication is clozapine, and it is primarily the recognition of the significant affinity of clozapine for 5-HT$_{2A}$ receptors which led to the 5-HT/DA hypothesis of schizophrenia. [Meltzer, 1989, 1992] However, clozapine also has significant affinity for D$_1$ receptors in PFC [Lundberg et al., 1989], leading to the suggestion that D$_1$ antagonists may be atypical antipsychotics. [see also Farde et al., 1989; Wadenberg et al., 1993] Clinical reports described above, utilizing selective 5-HT$_{2A}$ antagonists, have supported the argument in favor of the efficacy of these compounds in treating schizophrenia. [Axelsson et al., 1991; Bjerkenstedt, 1989; Ceulemans et al., 1985; Christensson and Bjork, 1990; Gelders, 1989] Therefore, we will now address data pertaining to the role of D$_1$ receptor antagonists in schizophrenia. [for additional review see Farde and Nordstrom, 1992; Gerlach and Hansen, 1992; Hietala et al., 1990]

Much of the data pertaining to D$_1$ receptor antagonism has utilized the compound SCH 23390. [Iorio et al., 1983] It is important to note, however, that SCH 23390 also acts as a 5-HT$_{2A}$ antagonist [Bischoff et al., 1986; see also Bijak and Smialowski, 1989; McQuade et al., 1988] and, possibly, as a 5-HT$_1$ agonist. [Skarsfeldt and Larsen, 1988] In the cells-per-track studies cited above, SCH 23390 had an atypical antipsychotic profile in one study.

[Goldstein and Litwin, 1988] However, Esposito and Bunney [1989] failed to demonstrate any evidence for antipsychotic efficacy of SCH 23390 in the same model. Acute and chronic SCH 23390 administration have also been shown to have no effect on 5-HT levels or the kinetic characteristics of 5-HT_{2A} receptors. [Gandolfi et al., 1988] These data, both electrophysiological and biochemical, stand in contrast to the above cited reports that SCH 23390 interacts with 5-HT_{2A} receptors, a property that should allow SCH 23390 to have direct effects on 5-HT systems. Based on these equivocal data, and the lack of clinical reports of selective D_1 antagonists being efficacious in the treatment of schizophrenia, it would appear that 5-HT_{2A} receptor antagonism is significantly more important than D_1 receptor blockade in engendering an atypical profile on antipsychotic drugs.

V. MDMA: DUAL ROLE FOR DA AND 5-HT

MDMA is a novel psychotropic agent related to both amphetamine and mescaline. [for general review see Whitaker-Azmitia and Peroutka, 1990] MDMA is typically recognized as a neurotoxin selective for 5-HT neurons. [Commins et al., 1987; Johnson et al., 1988; Schmidt, 1987; Stone et al., 1986] However, this compound also has significant effects on brain levels of DA and DA metabolites. [Commins et al., 1987; Johnson et al., 1988; Schmidt et al., 1987; Stone et al., 1986; Yamamoto and Spanos, 1988] Thus, MDMA could alter DA activity either directly and/or indirectly.

A. BIOCHEMICAL AND BEHAVIORAL STUDIES

Most literature on the neurotoxic effects of MDMA supports a mechanism in which MDMA-induced release of 5-HT causes DA release which then has a neurotoxic effect on 5-HT terminals in various brain regions. Stone et al. [1988] first demonstrated that endogenous DA was necessary for the neurotoxic effects of MDMA. [see also Brodkin et al., 1993] MDMA causes an acute release of both 5-HT and DA. [Fitzgerald and Reid, 1990; Gough et al., 1991; Hiramatsu et al., 1991; Nash and Brodkin, 1991] The DA precursor L-DOPA potentiates the neurotoxic effects of MDMA [Schmidt et al., 1991a], 5-HT_{2A} antagonists prevent the neurochemical/neurotoxic effects of MDMA [Schmidt et al., 1991b, 1992] and the disruption of MDMA's action by 5-HT_{2A} antagonists can be reversed by administration of L-DOPA. [Schmidt et al., 1991, 1992] In support of these data, 5-HT_{2A} receptor stimulation appears to be involved in MDMA-induced increases in DA synthesis. [Huang and Nichols, 1993] Although there is one study, utilizing *in vivo* voltammetry, which demonstrated a decrease in DA concentration in the neostriatum of rats following MDMA administration, this effect of MDMA was also reduced by prior depletion of 5-HT. [Gazzara et al., 1989] Thus, the neurochemical/neurotoxic effects of MDMA do appear to involve an interaction between 5-HT and DA.

There are conflicting data regarding the role of DA in the behavioral and discriminative stimulus effects of MDMA. Gold and colleagues published several studies in which they demonstrated that MDMA induced locomotor activation similar to DA agonists, that DA depletion reduced this MDMA-induced behavioral response and that MDMA shares discriminative stimulus properties with the DA agonists amphetamine and cocaine. [Gold and Koob, 1989; Gold et al., 1988, 1989] That MDMA shares discriminative stimulus properties with amphetamine was first demonstrated by Glennon and Young [1984] and has been replicated. [Oberlender and Nichols, 1988] However, Broadbent et al. [1989] reported that MDMA did not substitute for cocaine. Similarly, additional studies have either supported or argued against a role for endogenous DA in the locomotor stimulant effects of MDMA. [Callaway and Geyer, 1992; Callaway et al., 1990] Finally, MDMA has effects similar to DA agonists on the acoustic startle reflex and prepulse inhibition thereof, but the weight of evidence does not support a role for DA in this behavioral paradigm. [Kehne et al., 1992; Mansbach et al., 1989; Schmidt and Kehne, 1990]

In conclusion, numerous studies have demonstrated a role for endogenous DA in the neurochemical, neurotoxic and behavioral effects of MDMA. However, MDMA also has effects which are distinct from its direct influence on 5-HT systems within the nervous system. Thus, MDMA's unique profile appears to involve a complex interaction of direct actions on 5-HT and both direct and indirect actions of midbrain DA systems.

B. ELECTROPHYSIOLOGICAL STUDIES

MDMA has been shown to inhibit the firing rate of serotonergic neurons in a dose-dependent fashion and via an indirect effect mediated by the release of endogenous 5-HT. [Piercey et al., 1990; Sprouse et al., 1989, 1990] MDMA also dose-dependently inhibits the firing rate of DA neurons in the substantia nigra, with ED_{50} values of approximately 4 mg/kg, i.v. [Kelland et al., 1989d; Matthews et al., 1989] Complete inhibition of a majority of DA neurons was seen by Kelland et al. [1989d], but typically required very high doses (>30 mg/kg) and was not reliably reversed by haloperidol. The inhibition of DA neurons by MDMA appeared to involve both 5-HT and DA components, because depletion of either neurotransmitter caused a significant reduction in the inhibitory effect of MDMA, whereas depletion of both 5-HT and DA nearly eliminated the inhibitory effect of MDMA on DA neurons. [Kelland et al., 1989d] Although an additional study failed to observe MDMA-induced inhibition of DA neurons [Piercey et al., 1990], this study reports a maximal dose of only 3 mg/kg of MDMA tested on only 4 cells, both the dose and the sample size being too low to reliably detect MDMA-induced responses (based on the doses reported above).

In addition to the effects of MDMA on the activity of DA neurons, this compound appears to have modulatory effects on midbrain DA systems. MDMA enhanced the sensitivity of DA neurons to quinpirole, and eliminated the rate-dependent inhibition of DA neurons by both quinpirole and the D_1/D_2 agonist apomorphine. [Kelland et al., 1989d] Since MDMA elicits an acute release of 5-HT *in vivo*, and 5-HT modulates glutamate neurotransmission in the substantia nigra, MDMA could be hypothesized to block glutamate-induced excitatory modulation of DA neurons. Subsequently, DA cells might be more sensitive to either quinpirole or other inhibitory influences. This proposal that "turning up" the raphe-nigral 5-HT input would facilitate DA neuronal inhibition is the converse of Bunney and DeRiemer's [1982] suggestion regarding the ability of clonidine to influence the activity of nigral DA neurons. However, this cannot explain why no corresponding increase in the sensitivity to apomorphine was observed following MDMA. [Kelland et al., 1989d]

As described above, either repeated or single, high-dose administration of MDMA induces a long-lasting depletion of 5-HT throughout the brain [Commins et al., 1987; Schmidt, 1987], and pretreatment with the 5-HT-depleting agents 5,7-DHT and PCPA eliminates the rate-dependent responsiveness of DA neurons to quinpirole. [Kelland et al., 1990a] Thus, chronic treatment with MDMA would also be expected to eliminate the rate-dependent responsiveness of DA neurons to quinpirole and we have, in fact, observed this phenomenon. [Kelland et al., 1989d] It has been proposed that fast-firing DA cells possess fewer somatodendritic autoreceptors and that this explains the rate-dependent nature of indirect-acting DA agonist-induced inhibition of midbrain DA neurons. [Chiodo et al., 1984; White and Wang, 1984] The 5-HT depletion data suggest that, whereas somatodendritic autoreceptors may be involved in directly mediating the inhibition itself, afferent regulation by 5-HT is essential for the occurrence and/or maintenance of the rate-dependent nature of that inhibition. It is important to note, however, that acute MDMA pretreatment also eliminated the rate-dependent responsiveness of DA neurons. Since acute MDMA increases 5-HT release, this would appear to contradict the 5-HT depletion data. [Johnson et al., 1986; Schmidt et al., 1987] Perhaps some critical level of 5-HT must be maintained in order for this neurotransmitter to properly modulate the functioning of midbrain DA systems.

In a preliminary report, Ashby et al. [1993] have reported that MDMA also inhibits the firing rate of DA neurons in the VTA as well as nucleus accumbens neurons. The ED_{50} for MDMA-induced inhibition of VTA DA neurons was approximately the same as for DA neurons in the substantia nigra. [Kelland et al., 1989d; Matthews et al., 1989] In contrast to our results, however, the inhibition of DA neurons in the VTA seemed primarily dependent on DA release, whereas MDMA-induced inhibition of nigrostriatal DA neurons involved significant contributions from both DA and 5-HT pathways. [Ashby et al., 1993; Kelland et al., 1989d]

MDMA might also interact with DA in the mPFC. MDMA inhibits mPFC neurons. [Pan and Wang, 1991a,b] This MDMA-induced inhibition is blocked by pretreatment with a 5-HT uptake blocker but not by a DA uptake blocker, is also blocked by depletion of 5-HT by PCPA but not by depletion of DA by alpha-methyl-p-tyrosine, and the effects of PCPA can be reversed by restoring 5-HT content with the administration of 5-hydroxytryptophan. [Pan and Wang, 1991a,b] Thus, the inhibition of mPFC neurons by MDMA appears to exclusively involve the release of 5-HT (as opposed to a contribution by DA release in addition to 5-HT release, as in the substantia nigra). [Kelland et al., 1989d] Regardless of the mechanism, by suppressing the activity of these cells MDMA exerts a significant influence on the interaction between 5-HT and DA in this brain region.

VI. FURTHER EVIDENCE FOR INTERACTIONS BETWEEN 5-HT AND DA

In this section we will briefly address the major areas not specifically covered above in which interactions between 5-HT and DA have been identified. In the last section, we will touch on DA modulation of 5-HT. Although these data are not the topic of the current review, they deserve mention because of the general relationship between the two neurotransmitters, and because DA modulation of 5-HT systems will, of course, subsequently affect 5-HT modulation of DA systems.

A. 5-HT MODULATION OF DA RELEASE

There are conflicting data on the effects of 5-HT on DA release in DA terminal regions. 5-HT has been shown to increase DA release in the striatum and nucleus accumbens [Benloucif and Galloway, 1991; Benloucif et al., 1993; Blandina et al., 1989; Parsons and Justice, 1993], to increase DA metabolism without affecting release [DeSimoni et al., 1987] or to inhibit DA release. [Muramatsu et al., 1988] Both 5-HT_{2A} and, possibly, 5-HT_3 receptors appear to mediate 5-HT-induced increases in DA release, whereas 5-HT_{2A} receptor antagonists blocked the inhibition of DA release by 5-HT. [Blandina et al., 1989; Muramatsu et al., 1988; Parsons and Justice, 1993] Studies utilizing 5-HT agonists have more reliably demonstrated that 5-HT facilitates DA release, with contributions by 5-HT_1, 5-HT_{2A} and 5-HT_3 receptors. [Benloucif and Galloway, 1991; Benloucif et al., 1993; Chen et al., 1991; Devaud and Hollingsworth, 1991; Hamon et al., 1988; Jiang et al., 1990; Nissbrandt et al., 1992] Thus, the majority of data support the conclusion that 5-HT elicits the release of DA. This conclusion is also supported by data on the mechanism of action of MDMA-induced neurotoxicity (see above).

B. 5-HT INFLUENCE ON DA-MEDIATED BEHAVIOR

Several early studies demonstrated that injections of 5-HT directly into the nucleus accumbens reduced the locomotor stimulant effects of either DA or amphetamine [Carter and Pycock, 1978; Costall et al., 1976, 1979], whereas

5-HT depletion enhanced locomotion induced by DA agonists. [Costall et al., 1976, 1979; Lyness and Moore, 1981] These results are generally consistent with an inhibitory effect of 5-HT on DA function, in agreement with most electrophysiological studies (see above). More recent studies, utilizing 5-HT compounds selective for receptor subtypes, have yielded confusing results. Systemic administration of 5-HT_1 agonists has been shown to enhance locomotor activity. [Oberlander et al., 1987; Tricklebank et al., 1985; Young et al., 1993] However, there is conflicting evidence as to whether this locomotor activity is dependent on DA function. [Tricklebank et al., 1985; Young et al., 1993] When injected directly into the nucleus accumbens, individual 5-HT_1 receptor subtype selective compounds did not alter amphetamine-induced locomotion, but a combination of compounds did. [Layer et al., 1992] Finally, 5-HT_3 receptors in the nucleus accumbens appear to mediate a 5-HT enhancement of either amphetamine- or neurokinin-stimulated locomotor activity. [Costall et al., 1987; Hagan et al., 1987; Layer et al., 1992] Thus, the more recent studies utilizing 5-HT receptor subtype selective drugs generally disagree with the early studies utilizing 5-HT itself, i.e., 5-HT appears to inhibit DA-mediated behaviors but 5-HT drugs appear to stimulate DA-mediated behavior. At present, there is no obvious answer to this dilemma.

A series of electrophysiological studies were described above which suggested that 5-HT_{2A} antagonists have significant potential for treating schizophrenia, either alone or as adjuncts to conventional therapy, and that 5-HT_{2A} antagonism may be a critical property engendering an atypical antipsychotic profile. Another model used to predict the propensity for extrapyramidal side effects is to measure DA antagonist-induced catalepsy. It has now been reported that the 5-HT_{1A} agonist DPAT can reverse haloperidol-induced catalepsy in rats. [Broekkamp et al., 1988; Invernizzi et al., 1988] Based on these data, another type of atypical antipsychotic medication would be predicted if compounds are identified with a D_2 antagonist/5-HT_{1A} agonist profile.

C. 5-HT, COCAINE, AND OTHER DRUGS OF ABUSE

Recent data suggest that 5-HT systems might play an important role in the maintenance of illicit drug use. Cocaine is a DA agonist, acting primarily at the DA uptake site. However, cocaine also has effects on 5-HT neurons. Cocaine inhibits the electrophysiological activity of 5-HT neurons in the dorsal raphe. [Cunningham and Lakoski, 1988; Pan and Williams, 1989] Cocaine also inhibits the synthesis of both 5-HT and DA, most likely via autoregulatory pathways. [Baumann et al., 1993; Galloway, 1990] These direct actions on 5-HT systems by cocaine, as well as the many sites of interaction between 5-HT and DA described above, suggest that 5-HT could be a significant factor in the pharmacological actions of cocaine. In separate studies, a 5-HT_{1A} agonist and a 5-HT_3 antagonist were found to reduce self-administration of cocaine and cocaine-induced increases in extracellular DA levels, respectively. [McNeish et al., 1993; Peltier and Schenk, 1993] However, 5-HT$_3$ antagonists do not appear able to reduce the effects of cocaine on DA neurons. [Batsche et al., 1992] Thus, data suggesting that 5-HT plays a significant role in the central nervous system effects of cocaine remain equivocal.

The effects of 5-HT$_3$ antagonists on other reinforcing drugs are also unclear. These compounds reduce increases in extracellular DA levels induced by morphine, nicotine and ethanol. [Carboni et al., 1989; Imperato and Angelucci, 1989; Wozniak et al., 1990] However, 5-HT$_3$ antagonists have no effect on amphetamine-induced release of DA. [Carboni et al., 1989; Montgomery et al., 1993] Still, in support of a role for 5-HT in reinforcing mechanisms, indirect 5-HT agonists (various antidepressants) facilitate intracranial self-stimulation, but 5-HT$_3$ antagonists reduce the effects of nicotine on self-stimulation. [McCarter and Kokkinidis, 1988; Montgomery et al., 1993] Taken together, these data suggest that 5-HT has significant effects on reward processes, but at sites other than the DA synapse itself.

D. DA INFLUENCE ON 5-HT SYSTEMS

Lee and colleagues have conducted a series of studies on the effects of the mixed D$_1$/D$_2$ agonist apomorphine on the release and metabolism of 5-HT. Apomorphine causes increased 5-HT fluorescence in the dorsal raphe and increased 5-HT release in the neostriatum. [Lee, 1987; Lee and Geyer, 1982, 1983] These effects appear to be selective for the dorsal raphe, as opposed to the median raphe. [Lee and Geyer, 1982, 1983] Further evidence has shown that the effects of apomorphine on the dorsal raphe 5-HT system are indirect, since the effects of systemic injection of apomorphine can be blocked by depletion of DA, and direct injections of apomorphine into the substantia nigra, but not the dorsal raphe, elicit the same biochemical responses as systemic injection. [Lee and Geyer, 1984a,b] Neurotransmitter systems mediating the indirect effects of apomorphine on 5-HT neurons appear to include acetylcholine, GABA and DA (the latter via DA autoreceptors). [Chen et al., 1992a; Lee and Geyer, 1984a] Additional investigators have also demonstrated that DA and DA agonists increase the release of 5-HT. [Balfour and Iyaniwura, 1985; Ferre and Artigas, 1993; Kelly et al., 1985] Ferre and Artigas [1993] showed that the effects of apomorphine on 5-HT neurons appear to be mediated via D$_2$ receptors. This is important to note in light of evidence that D$_1$ receptor stimulation inhibits the release of 5-HT. [Benkirane et al., 1987]

VII. CONCLUSION

Despite the breadth of data described in this chapter, it is by no means an exhaustive coverage of the literature on interactions between 5-HT and DA. Thus, it is all the more surprising that the majority of reviews on either DA or

5-HT function tend to overlook the influence of these neurotransmitters on each other. Perhaps one reason is the difficulty in arriving at any distinct conclusions regarding the influence of 5-HT on DA systems, and vice versa. Generally, 5-HT has an inhibitory influence on the activity of DA neurons, in both direct and modulatory fashions. This conclusion is supported by the early behavioral literature. However, 5-HT causes DA release, and this constitutes part of the mechanism of action of MDMA-induced neurotoxicity. Resulting from the ability of 5-HT to enhance DA release, an important aspect of the 5-HT influence on DA cell activity is a modulation of dendritic DA release and subsequent autoreceptor-mediated inhibition of impulse generation by DA cells. 5-HT also significantly influences the activity of PFC pathways believed to be involved in schizophrenia, and the unique profile of atypical antipsychotics may be attributable to their 5-HT antagonist properties. Given the profound clinical significance of drug treatments for schizophrenia, and preliminary evidence that 5-HT antagonists may block drug-induced reinforcement (i.e., a treatment for the abuse of illicit drugs), continued research into the interactions between 5-HT and DA systems is clearly warranted.

ACKNOWLEDGEMENTS

The authors would like to thank Dr. Donna Z. Kelland for her helpful comments regarding this manuscript, and Ms. J. Rubin for technical assistance during the course of our research. The authors were supported by grants from the Tourette Syndrome Association (MDK) and PHS Grant MH41557 (LAC).

REFERENCES

Abi-Dargham, A., Laruelle, M., Lipska, B., Jaskiw, G. E., Wong, D. T., Robertson, D. W., Weinberger, D. R. and Kleinman, J. E., Serotonin 5-HT$_3$ receptors in schizophrenia: A postmortem study of the amygdala, *Brain Res.*, 616, 53, 1993.

Aghajanian, G. K. and Bunney, B. S., Dopaminergic and non-dopaminergic neurons of the substantia nigra: Differential responses to putative transmitters, in *Proceedings of the IX Congress of the College on International Neuropsychopharmacology*, Boissier, J. R., Hippius, H. and Pichot, P., Eds., Excerpta Medica, Amsterdam, 1974, pg. 444.

Aghajanian, G. K., Wang, R. Y. and Baraban, J., Serotonergic and nonserotonergic neurons of the dorsal raphe: Reciprocal changes in firing induced by peripheral nerve stimulation, *Brain Res.*, 153, 169, 1978.

Andreasen, N. C., Positive and negative symptoms: Historical and conceptual aspects, *Mod. Probl. Pharmacopsychiatry*, 24, 1, 1990.

Arborelius, L., Chergui, K., Murase, S., Nomikos, G. G., Hook, B. B., Chouvet, G., Hacksell, U. and Svensson, T. H., The 5-HT$_{1A}$ receptor selective ligands, (R)-8-OH-DPAT and (S)-UH-301, differentially affect the activity of midbrain dopamine neurons, *Naunyn-Schmiedeberg's Arch. Pharmacol.*, 347, 353, 1993a.

Arborelius, L., Nomikos, G. G., Hacksell, U. and Svensson, T. H., (R)-8-OH-DPAT preferentially increases dopamine release in rat medial prefrontal cortex, *Acta Physiol. Scand.*, 148, 465, 1993b.

Ashby, C. R., Jr., Edwards, E., Harkins, K. and Wang, R. Y., Characterization of 5-hydroxytryptamine₃ receptors in the medial prefrontal cortex: A microiontophoretic study, *Eur. J. Pharmacol.*, 173, 193, 1989a.

Ashby, C. R., Jr., Edwards, E., Harkins, K. L. and Wang, R. Y., Differential effect of typical and atypical antipsychotic drugs on the suppressant action of 2-methylserotonin on medial prefrontal cortical cells: A microiontophoretic study, *Eur. J. Pharmacol.*, 166, 583, 1989b.

Ashby, C. R., Jr., Jiang, L. H., Kasser, R. J. and Wang, R. Y., Electrophysiological characterization of 5-hydroxytryptamine₂ receptors in the rat medial prefrontal cortex, *J. Pharmacol. Exp. Ther.*, 252, 171, 1990a.

Ashby, C. R., Jr., Jiang, L. H. and Wang, R. Y., Chronic BRL 43694, a selective 5-HT₃ receptor antagonist, fails to alter the number of spontaneously active midbrain dopamine neurons, *Eur. J. Pharmacol.*, 175, 347, 1990b.

Ashby, C. R., Jr., Minabe, Y., Edwards, E. and Wang, R. Y., 5-HT₃-like receptors in the rat medial prefrontal cortex: An electrophysiological study, *Brain Res.*, 550, 181, 1991a.

Ashby, C. R., Jr., Minabe, Y., Edwards, E. and Wang, R. Y., Comparison of the effects of various typical and atypical antipsychotic drugs on the suppressant action of 2-methylserotonin on medial prefrontal cortical cells in the rat, *Synapse*, 8, 155, 1991b.

Ashby, C. R., Jr., Minabe, Y., Toor, A., Emori, K. and Wang, R. Y., The effect of (+), (-) and (±)-methylenedioxymethamphetamine (MDMA) on the activity of A10 dopamine (DA) neurons in the ventral tegmental area (VTA) and A10 target neurons in the nucleus accumbens (NAc), *Soc. Neurosci. Abs.*, 19, 828, 1993.

Ashby, C. R., Jr., Minabe, Y., Toor, A., Fishkin, L. D., Granoff, M. I. and Wang, R. Y., Effect produced by acute and chronic administration of the selective 5-HT₃ receptor antagonist BRL 46470 on the number of spontaneously active midbrain dopamine cells in the rat, *Drug Development Res.*, 31, 228, 1994.

Ashby, C. R., Jr. and Wang, R. Y., Effects of antipsychotic drugs on 5-HT₂ receptors in the medial prefrontal cortex: Microiontophoretic studies, *Brain Res.*, 506, 346, 1990.

Awouters, F., Niemegeers, C. J. E., Megens, A. A. H. P., Meert, T. F. and Janssen, P. A. J., Pharmacological profile of ritanserin: A very specific central serotonin S₂-antagonist, *Drug Devel. Res.*, 15, 61, 1988.

Axelsson, R., Nilsson, A., Christensson, E. and Bjork, A., Effects of amperozide in schizophrenia, *Psychopharmacol.*, 104, 287, 1991.

Balfour, D. J. K. and Iyaniwura, T. T., An investigation of amphetamine-induced release of 5-HT from rat hippocampal slices, *Eur. J. Pharmacol.*, 109, 395, 1985.

Bannon, M. J., Kelland, M. and Chiodo, L. A., Medial forebrain bundle stimulation or D-2 dopamine receptor activation increases preproenkephalin mRNA in rat striatum, *J. Neurochem.*, 52, 859, 1989.

Batsche, K., Granoff, M. I. and Wang, R. Y., 5-HT$_3$ receptor antagonists fail to block the suppressant effect of cocaine on the firing rate of A10 dopamine neurons in the rat, *Brain Res.*, 592, 273, 1992.

Baumann, M. H., Raley, T. J., Partilla, J. S. and Rothman, R. B., Biosynthesis of dopamine and serotonin in the rat brain after repeated cocaine injections: A microdissection mapping study, *Synapse*, 14, 40, 1993.

Benkirane, S., Arbilla, S. and Langer, S. Z., A functional response to D$_1$ dopamine receptor stimulation in the central nervous system: Inhibition of the release of [^{3}H]-serotonin from the rat substantia nigra, *Naunyn-Schmiedeberg's Arch. Pharmacol.*, 335, 502, 1987.

Benloucif, S. and Galloway, M. P., Facilitation of dopamine release in vivo by serotonin agonists: Studies with microdialysis, *Eur. J. Pharmacol.*, 200, 1, 1991.

Benloucif, S., Keegan, M. J. and Galloway, M. P., Serotonin facilitated dopamine release in vivo: Pharmacological characterization, *J. Pharmacol. Exp. Ther.*, 265, 373, 1993.

Berger, T. W., Kaul, S., Stricker, E. M. and Zigmond, M. J., Hyperinnervation of the striatum by dorsal raphe afferents after dopamine-depleting brain lesions in neonatal rats, *Brain Res.*, 336, 354, 1985.

Bijak, M. and Smialowski, A., Serotonin receptor blocking effect of SCH 23390, *Pharmacol. Biochem. Behav.*, 32, 585, 1989.

Bischoff, S., Heinrich, M., Sonntag, J. M. and Krauss, J., The D-1 dopamine receptor antagonist SCH 23390 also interacts potently with brain serotonin (5HT2) receptors, *Eur. J. Pharmacol.*, 129, 367, 1986.

Bjerkenstedt, L., Melperone in the treatment of schizophrenia, *Acta Psychiatr. Scand.*, 352, 35, 1989.

Blandina, P., Goldfarb, J., Craddock-Royal, B. and Green, J. P., Release of endogenous dopamine by stimulation of 5-hydroxytryptamine$_3$ receptors in rat striatum, *J. Pharmacol. Exp. Ther.*, 251, 803, 1989.

Blier, P. and De Montigny, C., Modification of 5-HT neurons properties by sustained administration of the 5-HT1A agonist gepirone: Electrophysiological studies in the rat brain, *Synapse*, 1, 470, 1987.

Blier, P., De Montigny, C. and Tardif, D., Short-term lithium treatment enhances responsiveness of postsynaptic 5-HT$_{1A}$ receptors without altering 5-HT autoreceptor sensitivity: An electrophysiological study in the rat brain, *Synapse*, 1, 225, 1987.

Broadbent, J., Michael, E. K. and Appel, J. B., Generalization of cocaine to the isomers of 3,4-methylenedioxyamphetamine and 3,4-

methylenedioxymethamphetamine: Effects of training dose, *Drug Devel. Res.*, 16, 443, 1989.

Brodkin, J., Malyala, A. and Nash, J. F., Effect of acute monoamine depletion on 3,4-methylenedioxymethamphetamine-induced neurotoxicity, *Pharmacol. Biochem. Behav.*, 45, 647, 1993.

Broekkamp, C. L. E., Oosterloo, S. K., Berendsen, H. H. G. and van Delft, A. M. L., Effect of metergoline, fenfluramine, and 8-OHDPAT on catalepsy induced by haloperidol or morphine, *Naunyn-Schmiedeberg's Arch. Pharmacol.*, 338, 191, 1988.

Bruno, J. P., Jackson, D., Zigmond, M. J. and Stricker, E. M., Effect of dopamine-depleting brain lesions in rat pups: Role of striatal serotonergic neurons in behavior, *Behav. Neurosci.*, 101, 806, 1987.

Bruno, J. P., Snyder, A. M. and Stricker, E. M., Effect of dopamine-depleting brain lesions on suckling and weaning in rats, *Behav. Neurosci.*, 98, 156, 1984.

Bunney, B. S. and DeRiemer, S., Effect of clonidine on dopaminergic neurons activity in the substantia nigra: Possible indirect mediation by noradrenergic regulation of the serotonergic raphe system, in *Gilles de la Tourette Syndrome*, Friedhoff, A. J. and Chase, T. N., Eds., Raven Press, New York, 1982, pg. 99.

Bunney, B. S. and Grace, A. A., Acute and chronic haloperidol treatment: Comparison of effects on nigral dopaminergic cell activity, *Life Sci.*, 23, 1715, 1978.

Bunney, B. S. and Sesack, S. R., Electrophysiological identification and pharmacological characterization of dopamine sensitive neurons in the rat prefrontal cortex, in *Neurophysiology of Dopaminergic Systems - Current Status and Clinical Perspectives*, Chiodo, L. A. and Freeman, A. S., Eds., Lakeshore Publishing Co., Grosse Pointe, 1987, chap. 4.

Bunney, B. S., Walters, J. R., Roth, R. H. and Aghajanian, G. K., Dopaminergic neurons: Effect of antipsychotic drugs and amphetamine on single cell activity, *J. Pharmacol. Exp. Ther.*, 185, 560, 1973.

Callaway, C. W. and Geyer, M. A., Stimulant effects of 3,4-methylenedioxymethamphetamine in the nucleus accumbens of rat, *Eur. J. Pharmacol.*, 214, 45, 1992.

Callaway, C. W., Wing, L. L. and Geyer, M. A., Serotonin release contributes to the locomotor stimulant effects of 3,4-methylenedioxymethamphetamine in rats, *J. Pharmacol. Exp. Ther.*, 254, 456, 1990.

Carboni, E., Acquas, E., Frau, R. and DiChiara, G., Differential inhibitory effects of a 5-HT3 antagonist on drug-induced stimulation of dopamine release, *Eur. J. Pharmacol.*, 164, 515, 1989.

Carlsson, A., The current status of the dopamine hypothesis of schizophrenia, *Neuropsychopharmacol.*, 1, 179, 1988.

Carpenter, M. B. and Sutin, J., *Human Neuroanatomy, 8th Ed.*, Williams &

Wilkins, Baltimore, 1983.

Carter, C. J. and Pycock, C. J., Differential effects of central serotonin manipulation on hyperactive and stereotyped behaviour, *Life Sci.*, 23, 953, 1978.

Ceulemans, D. L. S., Gelders, Y. G., Hoppenbrouwers, M.-L. J. A., Reyntjens, A. J. M. and Janssen, P. A. J., Effect of serotonin antagonism in schizophrenia: A pilot study with setoperone, *Psychopharmacol.*, 85, 329, 1985.

Chen, H. Y., Lin, Y. P. and Lee, E. H. Y., Cholinergic and GABAergic mediations of the effects of apomorphine on serotonin neurons, *Synapse*, 10, 34, 1992a.

Chen, J., Paredes, W., van Praag, H. M. and Gardner, E. L., Serotonin denervation enhances responsiveness of presynaptic dopamine efflux to acute clozapine in nucleus accumbens but not in caudate-putamen, *Brain Res.*, 582, 173, 1992b.

Chen, J., van Praag, H. M. and Gardner, E. L., Activation of 5-HT$_3$ receptor by 1-phenylbiguanide increases dopamine release in the rat nucleus accumbens, *Brain Res.*, 543, 354, 1991.

Chiodo, L. A., Dopamine-containing neurons in the mammalian central nervous system: Electrophysiology and pharmacology, *Neurosci. Biobehav. Rev.*, 12, 49, 1988.

Chiodo, L. A., Dopamine autoreceptor signal transduction in the DA cell body: A "current view," *Neurochem. Int.*, 20 (Suppl.), 81S, 1992.

Chiodo, L. A., Antelman, S. M., Caggiula, A. R. and Lineberry, C. G., Sensory stimuli alter the discharge rate of dopamine (DA) neurons: Evidence for two functional types of DA cells in the substantia nigra, *Brain Res.*, 189, 544, 1980.

Chiodo, L. A., Bannon, M. J., Grace, A. A., Roth, R. H. and Bunney, B. S., Evidence for the absence of impulse-regulating somatodendritic and synthesis-modulating nerve terminal autoreceptors on subpopulations of midbrain dopamine neurons, *Neurosci.*, 12, 1, 1984.

Chiodo, L. A. and Bunney, B. S., Typical and atypical neuroleptics: Differential effects of chronic administration on the activity of A9 and A10 midbrain dopaminergic neurons, *J. Neurosci.*, 3, 1607, 1983.

Chiodo, L. A. and Bunney, B. S., Possible mechanisms by which repeated clozapine administration differentially affects the activity of two subpopulations of midbrain dopamine neurons, *J. Neurosci.*, 5, 2539, 1985.

Chiodo, L. A., Caggiula, A. R., Antelman, S. M. and Lineberry, C. G., Reciprocal influences of activating and immobilizing stimuli on the activity of nigrostriatal dopamine neurons, *Brain Res.*, 176, 385, 1979.

Chiodo, L. A. and Kapatos, G., Mesencephalic neurons in primary culture: Immunohistochemistry and membrane physiology, in *Neurophysiology of Dopaminergic Systems - Current Status and Clinical Perspectives*, Chiodo, L. A. and Freeman, A. S., Eds., Lakeshore Publishing Co., Grosse Pointe, 1987, pg. 67.

Chiodo, L. A. and Kapatos, G., Membrane properties of identified mesencephalic dopamine neurons in primary dissociated cell culture, *Synapse*, 11, 294, 1992.

Christensson, E. and Bjork, A., Amperozide: A new pharmacological approach in the treatment of schizophrenia, *Pharmacol. Toxicol.*, 66 (Suppl. 1), 5, 1990.

Clark, D. and Chiodo, L. A., Electrophysiological and pharmacological characterization of identified nigrostriatal and mesoaccumbens dopamine neurons in the rat, *Synapse*, 2, 474, 1988.

Clark, D., Engberg, G., Pileblad, E., Svensson, T. H., Carlsson, A., Freeman, A. S. and Bunney, B. S., An electrophysiological analysis of the actions of the 3-PPP enantiomers on the nigrostriatal dopamine system, *Naunyn-Schmiedeberg's Arch. Pharmacol.*, 329, 344, 1985.

Commins, D. L., Vosmer, G., Virus, R. M., Woolverton, W. L., Schuster, C. R. and Seiden, L. S., Biochemical and histological evidence that methylenedioxymethamphetamine (MDMA) is toxic to neurons in the rat brain, *J. Pharmacol. Exp. Ther.*, 241, 338, 1987.

Costall, B., Domeney, A. M., Naylor, R. J. and Tyers, M. B., Effects of the 5-HT$_3$ receptor antagonist, GR38032F, on raised dopaminergic activity in the mesolimbic system of the rat and marmoset brain, *Br. J. Pharmac.*, 92, 881, 1987.

Costall, B., Hui, S.-C. G. and Naylor, R. J., The importance of serotonergic mechanisms for the induction of hyperactivity by amphetamine and its antagonism by intra-accumbens (3,4-dihydroxy-phenylamino)-2-imidazoline (DPI), *Neuropharmacol.*, 18, 605, 1979.

Costall, B., Naylor, R. J., Marsden, C. D. and Pycock, C. J., Serotoninergic modulation of the dopamine response from the nucleus accumbens, *J. Pharm. Pharmac.*, 28, 523, 1976.

Cox, R. F., Meller, E. and Waszczak, B. L., Electrophysiological evidence for a large receptor reserve for inhibition of dorsal raphe neuronal firing by 5-HT$_{1A}$ agonists, *Synapse*, 14, 297, 1993.

Crossman, A. R., Walker, R. J. and Woodruff, G. N., Pharmacological studies on single neurones in the substantia nigra of the rat, *Br. J. Pharmacol.*, 51, 137P, 1974.

Cunningham, K. A. and Lakoski, J. M., Electrophysiological effects of cocaine and procaine on dorsal raphe serotonin neurons, *Eur. J. Pharmacol.*, 148, 457, 1988.

Dahlstrom, A. and Fuxe, K., Evidence for the existence of monoamine-containing neurons in the central nervous system. 1. Demonstration of monoamines in the cell bodies of brain stem neurons, *Acta Physiol. Scand.*, 232 (Suppl. 62), 1, 1964.

Davies, J. and Tongroach, P., Neuropharmacological studies on the nigro-striatal and raphe-striatal system in the rat, *Eur. J. Pharmacol.*, 51, 91, 1978.

De Montigny, C. and Aghajanian, G. K., Preferential action of 5-methoxytryptamine and 5-methoxydimethyltryptamine on presynaptic serotonin receptors: A comparative iontophoretic study with LSD and serotonin, *Neuropharmacol.*, 16, 811, 1977.

Deutch, A. Y., Moghaddam, B., Innis, R. B., Krystal, J. H, Aghajanian, G. K., Bunney, B. S. and Charney, D. S., Mechanisms of action of atypical antipsychotic drugs: Implications for novel therapeutic strategies for schizophrenia, *Schizophr. Res.*, 4, 121, 1991.

Devaud, L. L. and Hollingsworth, E. B., Effects of the 5-HT$_2$ receptor antagonist, ritanserin, on biogenic amines in the rat nucleus accumbens, *Eur. J. Pharmacol.*, 192, 427, 1991.

DeSimoni, M. G., DalToso, G., Fodritto, F., Sokola, A. and Algeri, S., Modulation of striatal dopamine metabolism by the activity of dorsal raphe serotonergic afferences, *Brain Res.*, 411, 81, 1987.

Dray, A., Gonye, T. J., Oakley, N. R. and Tanner, T., Evidence for the existence of a raphe projection to the substantia nigra in rat, *Brain Res.*, 113, 45, 1976.

Drescher, K. and Hetey, L., Influence of antipsychotics and serotonin antagonists on presynaptic receptors modulating the release of serotonin in synaptosomes of the nucleus accumbens of rats, *Neuropharmacol.*, 27, 31, 1988.

Esposito, E. and Bunney, B. S., The effect of acute and chronic treatment with SCH 23390 on the spontaneous activity of midbrain dopamine neurons, *Eur. J. Pharmacol.*, 162, 109, 1989.

Farde, L. and Nordstrom, A.-L., PET analysis indicates atypical central dopamine receptor occupancy in clozapine-treated patients, *Br. J. Psychiat.*, 160 (Suppl. 17), 30, 1992.

Farde, L., Wiesel, F. A., Nordstrom, A.-L. and Sedvall, G., D1- and D2-dopamine receptor occupancy during treatment with conventional and atypical neuroleptics, *Psychopharmacol.*, 99, S28, 1989.

Feger, J., Jacquemin, J. and Ohye, C., Peripheral excitatory input to substantia nigra, *Exp. Neurol.*, 59, 351, 1978.

Ferre, S. and Artigas, F., Dopamine D$_2$ receptor-mediated regulation of serotonin extracellular concentration in the dorsal raphe nucleus of freely moving rats, *J. Neurochem.*, 61, 772, 1993.

Fibiger, H. C. and Miller, J. J., An anatomical and electrophysiological investigation of the serotonergic projection from the dorsal raphe nucleus to the substantia nigra in the rat, *Neurosci.*, 2, 975, 1977.

Fischette, C. T., Nock, B. and Renner, K., Effects of 5,7-dihydroxytryptamine on serotonin$_1$ and serotonin$_2$ receptors throughout the rat central nervous system using quantitative autoradiography, *Brain Res.*, 421, 263, 1987.

Fitzgerald, J. L. and Reid, J. J., Effects of methylenedioxymethamphetamine on the release of monoamines from rat brain slices, *Eur. J. Pharmacol.*, 191,

217, 1990.

Galloway, M. P., Regulation of dopamine and serotonin synthesis by acute administration of cocaine, *Synapse*, 6, 63, 1990.

Gandolfi, O., Roncada, P. and Dall'Olio, R., Single or repeated administrations of SCH 23390 fail to affect serotonergic neurotransmission, *Neurosci. Lett.*, 92, 192, 1988.

Gariano, R. F. and Groves, P. M., Burst firing induced in midbrain dopamine neurons by stimulation of the medial prefrontal and anterior cingulate cortices, *Brain Res.*, 462, 194, 1988.

Gazzara, R. A., Takeda, H., Cho, A. K. and Howard, S. G., Inhibition of dopamine release by methylenedioxymethamphetamine is mediated by serotonin, *Eur. J. Pharmacol.*, 168, 209, 1989.

Gelders, Y. G., Thymosthenic agents, a novel approach in the treatment of schizophrenia, *Br. J. Psychiat.*, 155 (Suppl. 5), 33, 1989.

Gerlach, J. and Hansen, L., Clozapine and D1/D2 antagonism in extrapyramidal functions, *Br. J. Psychiat.*, 160 (Suppl. 17), 34, 1992.

Glennon, R. A., Serotonin receptors: Clinical implications, *Neurosci. Biobehav. Rev.*, 14, 35, 1990.

Glennon, R. A. and Young, R., Further investigation of the discriminative stimulus properties of MDA, *Pharmacol. Biochem. Behav.*, 20, 501, 1984.

Gold, L. H. and Koob, G. F., MDMA produces stimulant-like conditioned locomotor activity, *Psychopharmacol.*, 99, 352, 1989.

Gold, L. H., Hubner, C. B. and Koob, G. F., A role for the mesolimbic dopamine system in the psychostimulant actions of MDMA, *Psychopharmacol.*, 99, 40, 1989.

Gold, L. H., Koob, G. F. and Geyer, M. A., Stimulant and hallucinogenic behavioral profiles of 3,4-methylenedioxymethamphetamine and N-ethyl-3,4-methylenedioxyamphetamine in rats, *J. Pharmacol. Exp. Ther.*, 247, 547, 1988.

Goldstein, J. M. and Litwin, L. C., Spontaneous activity in A9 and A10 dopamine neurons after acute and chronic administration of the selective dopamine D-1 receptor antagonist SCH 23390, *Eur. J. Pharmacol.*, 155, 175, 1988.

Goldstein, J. M., Litwin, L. C., Sutton, E. B. and Malick, J. B., Effects of ICI 169,369, a selective serotonin$_2$ antagonist, in electrophysiological tests predictive of antipsychotic activity, *J. Pharmacol. Exp. Ther.*, 249, 673, 1989.

Gonon, F. G. and Buda, M. J., Regulation of dopamine release by impulse flow and by autoreceptors as studied by in vivo voltammetry in the rat striatum, *Neuroscience*, 14, 765, 1985.

Gough, B., Ali, S. F., Slikker, Jr., W. and Holson, R. R., Acute effects of 3,4-methylenedioxymethamphetamine (MDMA) on monoamines in rat caudate, *Pharmacol. Biochem. Behav.*, 39, 619, 1991.

Grace, A. A., Phasic versus tonic dopamine release and the modulation of dopamine system responsivity: A hypothesis for the etiology of schizophrenia, *Neuroscience*, 41, 1, 1991.

Grace, A. A. and Bunney, B. S., The control of firing pattern in nigral dopamine neurons: Burst firing, *J. Neurosci.*, 4, 2877, 1984.

Grace, A. A. and Bunney, B. S., Dopamine, in *Neurotransmitter Actions in the Vertebrate Nervous System*, Rogawski, M. A. and Barker, J. L., Eds., Plenum Publishing Corp., New York, 1985, chap. 9.

Grace, A. A., Hommer, D. W. and Bunney, B. S., Peripheral and striatal influences on nigral dopamine cells: Mediation by reticulata neurons, *Brain Res. Bull.*, 2 (Suppl. 5), 105, 1980.

Grace, A. A. and Onn, S. P., Morphology and electrophysiological properties of immunocytochemically identified rat dopamine neurons recorded in vitro, *J. Neurosci.*, 9, 3463, 1989.

Greenshaw, A. J., Behavioural pharmacology of 5-HT$_3$ receptor antagonists: A critical update on therapeutic potential, *TIPS*, 14, 265, 1993.

Grossman, C. J., Kilpatrick, G. J. and Bunce, K. T., Development of a radioligand binding assay for 5-HT$_4$ receptors in guinea-pig and rat brain, *Br. J. Pharmacol.*, 109, 618, 1993.

Hagan, R. M., Butler, A., Hill, J. M., Jordan, C. C., Ireland, S. J. and Tyers, M. B., Effect of the 5-HT$_3$ receptor antagonist, GR38032F, on responses to injection of a neurokinin agonist into the ventral tegmental area of the rat brain, *Eur. J. Pharmacol.*, 138, 303, 1987.

Hamon, M., Fattaccini, C.-M., Adrien, J., Gallissot, M.-C., Martin, P. and Gozlan, H., Alterations of central serotonin and dopamine turnover in rats treated with ipsapirone and other 5-hydroxytryptamine $_{1A}$ agonists with potential anxiolytic properties, *J. Pharmacol. Exp. Ther.*, 246, 745, 1988.

Hashimoto, T., Nishino, N., Nakai, H. and Tanaka, C., Increase in serotonin 5-HT$_{1A}$ receptors in prefrontal and temporal cortices of brains from patients with chronic schizophrenia, *Life Sci.*, 48, 355, 1991.

Herve, D., Pickel, V. M., Joh, T. H. and Beaudet, A., Serotonin axon terminals in the ventral tegmental area of the rat: Fine structure and synaptic input to dopaminergic neurons, *Brain Res.*, 435, 71, 1987.

Herve, D., Simon, H., Blanc, G., Lisoprawski, A., Le Moal, M., Glowinski, J. and Tassin, J. P., Increased utilisation of dopamine in the nucleus accumbens but not in the cerebral cortex after dorsal raphe lesion in the rat, *Neurosci. Lett.*, 15, 127, 1979.

Heym, J., Trulson, M. E. and Jacobs, B. L., Raphe unit activity in freely moving cats: Effects of phasic auditory and visual stimuli, *Brain Res.*, 232, 29, 1982.

Hietala, J., Lappalainen, J., Koulu, M. and Syvalahti, E., Dopamine D$_1$ receptor antagonism in schizophrenia: Is there reduced risk of extrapyramidal side-effects?, *TIPS*, 11, 406, 1990.

Hiramatsu, M., DiStefano, E., Chang, A. S. and Cho, A. K., A pharmacokinetic analysis of 3,4-methylenedioxymethamphetamine effects on monoamine concentrations in brain dialysates, *Eur. J. Pharmacol.*, 204, 135, 1991.

Hommer, D. W. and Bunney, B. S., Effects of sensory stimuli on the activity of dopaminergic neurons: Involvement of non-dopaminergic nigral neurons and striato-nigral pathways, *Life Sci.*, 27, 377, 1980.

Huang, X. and Nichols, D. E., 5-HT$_2$ receptor-mediated potentiation of dopamine synthesis and central serotonergic deficits, *Eur. J. Pharmacol.*, 238, 291, 1993.

Imperato, A. and Angelucci, L., 5-HT$_3$ receptors control dopamine release in the nucleus accumbens of freely moving rats, *Neurosci. Lett.*, 101, 214, 1989.

Invernizzi, R. W., Cervo, L. and Samanin, R., 8-Hydroxy-2-(di-N-propylamino) tetralin, a selective serotonin$_{1A}$ receptor agonist, blocks haloperidol-induced catalepsy by an action on raphe nuclei medianus and dorsalis, *Neuropharmacol.*, 27, 515, 1988.

Iorio, L. C., Barnett, A., Leitz, F. H., Houser, V. P. and Korduba, C. A., SCH 23390, a potential benzazepine antipsychotic with unique interactions on dopaminergic systems, *J. Pharmacol. Exp. Ther.*, 226, 462, 1983.

Jackson, D. and Abercrombie, E. D., In vivo neurochemical evaluation of striatal serotonergic hyperinnervation in rats depleted of dopamine at infancy, *J. Neurochem.*, 58, 890, 1992.

Jacobs, B. L., Fornal, C. A. and Wilkinson, L. O., Neurophysiological and neurochemical studies of brain serotonergic neurons in behaving animals, in *The Neuropharmacology of Serotonin*, Whitaker-Azmitia, P. M. and Peroutka, S. J., Eds., *Ann. N.Y. Acad. Sci.*, Vol. 600, New York, 1990, pg. 260.

Jiang, L. H., Ashby, Jr., C. R., Kasser, R. J. and Wang, R. Y., The effect of intraventricular administration of the 5-HT$_3$ receptor agonist 2-methylserotonin on the release of dopamine in the nucleus accumbens: An in vivo chronocoulometric study, *Brain Res.*, 513, 156, 1990.

Johnson, M. P., Hoffman, A. J. and Nichols, D. E., Effects of the enantiomers of MDA, MDMA and related analogues on [^{3}H]dopamine release from superfused rat brain slices, *Eur. J. Pharmacol.*, 132, 269, 1986.

Johnson, M., Letter, A. A., Merchant, K., Hanson, G. R. and Gibb, J. W., Effects of 3,4-methylenedioxymethamphetamine isomers on central serotonergic, dopaminergic and nigral neurotensin systems of the rat, *J. Pharmacol. Exp. Ther.*, 244, 977, 1988.

Johnson, S. W., Mercuri, N. B. and North, R. A., 5-Hydroxytryptamine$_{1B}$ receptors block the GABA$_B$ synaptic potential in rat dopamine neurons, *J. Neurosci.*, 12, 2000, 1992.

Kalivas, P. W., Neurotransmitter regulation of dopamine neurons in the ventral tegmental area, *Brain Res. Rev.*, 18, 75, 1993.

Kehne, J. H., McCloskey, T. C., Taylor, V. L., Black, C. K., Fadayel, G. M. and Schmidt, C. J., Effects of the serotonin releasers 3,4-methylenedioxymethamphetamine (MDMA), 4-chloroamphetamine (PCA) and fenfluramine on acoustic and tactile startle reflexes in rats, *J. Pharmacol. Exp. Ther.*, 260, 78, 1992.

Kelland, M. D., Chiodo, L. A. and Freeman, A. S., Use of ketamine in electrophysiological studies of midbrain dopamine neurons, in *Status of Ketamine in Anesthesiology*, Domino, E. F., Ed., NPP Books, Ann Arbor, 1989a, pg. 181.

Kelland, M. D., Freeman, A. S. and Chiodo, L. A., SKF 38393 alters the rate-dependent D2-mediated inhibition of nigrostriatal but not mesoaccumbens dopamine neurons, *Synapse*, 2, 416, 1988.

Kelland, M. D., Freeman, A. S. and Chiodo, L. A., D1 receptor activation enhances sciatic nerve stimulation-induced inhibition of nigrostriatal dopamine neurons, *Synapse*, 3, 339, 1989b.

Kelland, M. D., Freeman, A. S. and Chiodo, L. A., Chloral hydrate anesthesia alters the responsiveness of identified midbrain dopamine neurons to dopamine agonist administration, *Synapse*, 3, 30, 1989c.

Kelland, M. D., Freeman, A. S. and Chiodo, L. A., ($\pm$)-3,4-Methylenedioxymethamphetamine-induced changes in the basal activity and pharmacological responsiveness of nigrostriatal dopamine neurons, *Eur. J. Pharmacol.*, 169, 11, 1989d.

Kelland, M. D., Freeman, A. S. and Chiodo, L. A., Serotonergic afferent regulation of the basic physiology and pharmacological responsiveness of nigrostriatal dopamine neurons, *J. Pharmacol. Exp. Ther.*, 253, 803, 1990a.

Kelland, M. D., Freeman, A. S., LeWitt, P. A. and Chiodo, L. A., Effects of (+)-4-propyl-9-hydroxynaphthoxazine on midbrain dopamine neurons: An electrophysiological study, *J. Pharmacol. Exp. Ther.*, 255, 276, 1990b.

Kelland, M. D., Freeman, A. S., Rubin, J. and Chiodo, L. A., Ascending afferent regulation of rat midbrain dopamine neurons, *Brain Res. Bull.*, 31, 539, 1993.

Kelland, M. D., Pitts, D. K., Freeman, A. S. and Chiodo, L. A., Repeated SKF 38393 and nigrostriatal system neuronal responsiveness: Functional down-regulation is followed by up-regulation after withdrawal, *Naunyn-Schmiedeberg's Arch. Pharmacol.*, 343, 447, 1991.

Kelly, E., Jenner, P. and Marsden, C. D., The effects of dopamine and dopamine agonists on the release of ^{3}H-GABA and ^{3}H-5-HT from rat nigral slices, *Biochem. Pharmacol.*, 34, 2655, 1985.

Kostowski, W., Giacalone, E., Garattini, S. and Valzelli, L., Electrical stimulation of midbrain raphe: Biochemical, behavioral and bioelectrical effects, *Eur. J. Pharmacol.*, 7, 170, 1969.

Layer, R. T., Uretsky, N. J. and Wallace, L. J., Effect of serotonergic agonists in the nucleus accumbens on d-amphetamine-stimulated locomotion, *Life Sci.*, 50, 813, 1992.

Lee, E. H. Y., Additive effects of apomorphine and clonidine on serotonin neurons in the dorsal raphe, *Life Sci.*, 40, 635, 1987.

Lee, E. H. Y. and Geyer, M. A., Selective effects of apomorphine on dorsal raphe neurons: A cytofluorimetric study, *Brain Res. Bull.*, 9, 719, 1982.

Lee, E. H. Y. and Geyer, M. A., Similarities of the effects of apomorphine and 3-PPP on serotonin neurons, *Eur. J. Pharmacol.*, 94, 297, 1983.

Lee, E. H. Y. and Geyer, M. A., Indirect effects of apomorphine on serotoninergic neurons in rats, *Neurosci.*, 11, 437, 1984a.

Lee, E. H. Y. and Geyer, M. A., Dopamine autoreceptor mediation of the effects of apomorphine on serotonin neurons, *Pharmacol. Biochem. Behav.*, 21, 301, 1984b.

Llinas, R., Greenfield, S. A. and Jahnsen, H., Electrophysiology of pars compacta cells in the in vitro substantia nigra - A possible mechanism for dendritic release, *Brain Res.*, 294, 127, 1984.

Lum, J. T. and Piercey, M. F., Electrophysiological evidence that spiperone is an antagonist of 5-HT_{1A} receptors in the dorsal raphe nucleus, *Eur. J. Pharmacol.*, 149, 9, 1988.

Lundberg, T., Lindstrom, L.H., Hartvig, P., Eckernas, S.-A., Ekblom, B., Lundqvist, H., Fasth, K.J., Gullberg, P. and Langstrom, B., Striatal and frontal cortex binding of 11-C-labelled clozapine visualized by positron emission tomography (PET) in drug-free schizophrenics and healthy volunteers, *Psychopharmacol.*, 99, 8, 1989.

Luthman, J., Bolioli, B., Tsutsumi, T., Verhofstad, A. and Jonsson, G., Sprouting of striatal serotonin nerve terminals following selective lesions of nigro-striatal dopamine neurons in neonatal rat, *Brain Res. Bull.*, 19, 269, 1987.

Lynch, M. R., Schizophrenia and the D1 receptor: Focus on negative symptoms, *Prog. Neuro-Psychopharmacol. Biol. Psychiat.*, 16, 797, 1992.

Lyness, W. H. and Moore, K. E., Destruction of 5-hydroxytryptaminergic neurons and the dynamics of dopamine in nucleus accumbens septi and other forebrain regions of the rat, *Neuropharmacol.*, 20, 327, 1981.

Mansbach, R. S., Braff, D. L. and Geyer, M. A., Prepulse inhibition of the acoustic startle response is disrupted by N-ethyl-3,4-methylenedioxyamphetamine (MDEA) in the rat, *Eur. J. Pharmacol.*, 167, 49, 1989.

Mantz, J., Thierry, A. M. and Glowinski, J., Effect of noxious tail pinch on the discharge rate of mesocortical and mesolimbic dopamine neurons: Selective activation of the mesocortical system, *Brain Res.*, 476, 377, 1989.

Marcinkiewicz, M., Verge, D., Gozlan, H., Pichat, L. and Hamon, M. Autoradiographic evidence for the heterogeneity of 5-HT_1 sites in the rat brain, *Brain Res.*, 291, 159, 1984.

Matsubara, S., Matsubara, R., Kusumi, I., Koyama, T. and Yamashita, I., Dopamine D_1, D_2 and $Serotonin_2$ receptor occupation by typical and atypical antipsychotic drugs in vitro, *J. Pharmacol. Exp. Ther.*, 265, 498, 1993.

Matsubara, S. and Meltzer, H. Y., Effect of typical and atypical antipsychotic drugs on 5-HT_2 receptor density in rat cerebral cortex, *Life Sci.*, 45, 1397, 1989.

Matthews, R. T., Champney, T. H. and Frye, G. D., Effects of ($\pm$)3,4-Methylenedioxymethamphetamine (MDMA) on brain dopaminergic activity in rats, *Pharmacol. Biochem. Behav.*, 33, 741, 1989.

McCarter, B. D. and Kikkinidis, L., The effects of long-term administration of antidepressant drugs on intracranial self-stimulation responding in rats, *Pharmacol. Biochem. Behav.*, 31, 243, 1988.

McNeish, C. S., Svingos, A. L., Hitzemann, R. and Strecker, R. E., The 5-HT$_3$ antagonist zacopride attenuates cocaine-induced increases in extracellular dopamine in rat nucleus accumbens, *Pharmacol. Biochem. Behav.*, 45, 759, 1993.

McQuade, R. D., Ford, D., Duffy, R. A., Chipkin, R. E., Iorio, L. C. and Barnett, A., Serotonergic component of SCH 23390: In vitro and in vivo binding analyses, *Life Sci.*, 43, 1861, 1988.

Meltzer, H. Y., Clinical studies on the mechanism of action of clozapine: The dopamine-serotonin hypothesis of schizophrenia, *Psychopharmacology*, 99, S18, 1989.

Meltzer, H. Y., The importance of serotonin-dopamine interactions in the action of clozapine, *Br. J. Psychiat.*, 160 (Suppl. 17), 22, 1992.

Meltzer, H. Y., Matsubara, S. and Lee, J.-C., The ratios of serotonin$_2$ and dopamine$_2$ affinities differentiate atypical and typical antipsychotic drugs, *Psychopharmacol. Bull.*, 25, 390, 1989a.

Meltzer, H. Y., Matsubara, S. and Lee, J.-C., Classification of typical and atypical antipsychotic drugs on the basis of dopamine D-1, D-2 and serotonin$_2$ pK$_i$ values, *J. Pharmacol. Exp. Ther.*, 251, 238, 1989b.

Minabe, Y., Ashby, C. R., Jr., Schwartz, J. E. and Wang, R. Y., The 5-HT$_3$ receptor antagonists LY 277359 and granisetron potentiate the suppressant action of apomorphine on the basal firing rate of ventral tegmental dopamine cells, *Eur. J. Pharmacol.*, 209, 143, 1991a.

Minabe, Y., Ashby, C. R., Jr. and Wang, R. Y., The effect of acute and chronic LY 277359, a selective 5-HT$_3$ receptor antagonist, on the number of spontaneously active midbrain dopamine neurons, *Eur. J. Pharmacol.*, 209, 151, 1991b.

Mita, T., Hanada, S., Nishino, N., Kuno, T., Nakai, H., Yamadori, T., Mizoi, Y. and Tanaka, C., Decreased serotonin S$_2$ and increased dopamine D$_2$ receptors in chronic schizophrenics, *Biol. Psychiat.*, 21, 1407, 1986.

Montgomery, A. M. J., Rose, I. C. and Herberg, L. J., The effect of a 5-HT$_3$ receptor antagonist, ondansetron, on brain stimulation reward, and its interaction with direct and indirect stimulants of central dopaminergic transmission, *J. Neural Transm.*, 91, 1, 1993.

Muramatsu, M., Tamaki-Ohashi, J., Usuki, C., Araki, H., Chaki, S. and Aihara, H., 5-HT$_2$ antagonists and minaprine block the 5-HT-induced inhibition of dopamine release from rat brain striatal slices, *Eur. J. Pharmacol.*, 153, 89, 1988.

Nash, J. F. and Brodkin, J., Microdialysis studies on 3,4-methylenedioxymethamphetamine-induced dopamine release: Effect of dopamine uptake inhibitors, *J. Pharmacol. Exp. Ther.*, 259, 820, 1991.

Nedergaard, S., Bolam, J. P. and Greenfield, S. A., Facilitation of a dendritic calcium conductance by 5-hydroxytryptamine in the substantia nigra, *Nature*, 333, 174, 1988.

Nissbrandt, H., Waters, N. and Hjorth, S., The influence of serotoninergic drugs on dopaminergic neurotransmission in rat substantia nigra, striatum and limbic forebrain in vivo, *Naunyn-Schmiedeberg's Arch. Pharmacol.*, 346, 12, 1992.

Oberlander, C., Demassey, Y., Verdu, A., Van de Velde, D. and Bardelay, C., Tolerance to the serotonin 5-HT$_1$ agonist RU 24969 and effects on dopaminergic behaviour, *Eur. J. Pharmacol.*, 139, 205, 1987.

Oberlender, R. and Nichols, D. E., Drug discrimination studies with MDMA and amphetamine, *Psychopharmacol.*, 95, 71, 1988.

Palacios, J. M., Waeber, C., Hoyer, D. and Mengod, G., Distribution of serotonin receptors, in *The Neuropharmacology of Serotonin*, Whitaker-Azmitia, P. M. and Peroutka, S. J., Eds., *Ann. N.Y. Acad. Sci.*, Vol. 600, New York, 1990, pg. 36.

Pan, H. S. and Wang, R. Y., MDMA: Further evidence that its action in the medial prefrontal cortex is mediated by the serotonergic system, *Brain Res.*, 539, 332, 1991a.

Pan, H. S. and Wang, R. Y., The action of (±)-MDMA on medial prefrontal cortical neurons is mediated through the serotonergic system, *Brain Res.*, 543, 56, 1991b.

Pan, Z. Z. and Williams, J. T., Differential actions of cocaine and amphetamine on dorsal raphe neurons in vitro, *J. Pharmacol. Exp. Ther.*, 251, 56, 1989.

Parent, A., Descarries, L. and Beaudet, A., Organization of ascending serotonin systems in the adult rat brain. A radioautographic study after intraventricular administration of [^{3}H]5-hydroxytryptamine, *Neuroscience*, 6, 115, 1981.

Parsons, L. H. and Justice, Jr., J. B., Perfusate serotonin increases extracellular dopamine in the nucleus accumbens as measured by in vivo microdialysis, *Brain Res.*, 606, 195, 1993.

Pazos, A. and Palacios, J. M., Quantitative autoradiographic mapping of serotonin receptors in the rat brain. I. Serotonin-1 receptors, *Brain Res.*, 346, 205, 1985.

Pehek, E. A., Meltzer, H. Y. and Yamamoto, B. K., The atypical antipsychotic drug amperozide enhances rat cortical and striatal dopamine efflux, *Eur. J. Pharmacol.*, 240, 107, 1993.

Peltier, R. and Schenk, S., Effects of serotonergic manipulations on cocaine self-administration in rats, *Psychopharmacol.*, 110, 390, 1993.

Peroutka, S. J., 5-Hydroxytryptamine receptor subtypes, *Annu. Rev. Neurosci.*, 11, 45, 1988.

Piercey, M. F., Lum, J. T. and Palmer, J. R., Effects of MDMA ("ecstasy") on firing rates of serotonergic, dopaminergic, and noradrenergic neurons in the rat, *Brain Res.*, 526, 203, 1990.

Prisco, S., Pessia, M., Ceci, A., Borsini, F. and Esposito, E., Chronic treatment with DAU 6215, a new 5-HT$_3$ receptor antagonist, causes a selective decrease in the number of spontaneously active dopaminergic neurons in the rat ventral tegmental area, *Eur. J. Pharmacol.*, 214, 13, 1992.

Radja, F., Descarries, L., Dewar, K. M. and Reader, T. A., Serotonin 5-HT$_1$ and 5-HT$_2$ receptors in adult rat brain after neonatal destruction of nigrostriatal dopamine neurons: A quantitative autoradiographic study, *Brain Res.*, 606, 273, 1993.

Raine, A., Sheard, C., Reynolds, G.P. and Lencz, T., Pre-frontal structural and functional deficits associated with individual differences in schizotypal personality, *Schizophrenia Res.*, 7, 237, 1992.

Rasmussen, K., Stockton, M. E. and Czachura, J. F., The 5-HT$_3$ receptor antagonist zatosetron decreases the number of spontaneously active A10 dopamine neurons, *Eur. J. Pharmacol.*, 205, 113, 1991.

Rasmussen, K., Strecker, R. E. and Jacobs, B. L., Single unit response of noradrenergic, serotonergic and dopaminergic neurons in freely moving cats to simple sensory stimuli, *Brain Res.*, 369, 336, 1986.

Rech, R. H. and Gudelsky, G. A., Eds., *5-HT Agonists as Psychoactive Drugs*, NPP Books, Ann Arbor, 1988.

Saller, C. F., Czupryna, M. J. and Salama, A. I., 5-HT$_2$ receptor blockade by ICI 169,369 and other 5-HT$_2$ antagonists modulates the effects of D-2 dopamine receptor blockade, *J. Pharmacol. Exp. Ther.*, 253, 1162, 1990.

Schmidt, C. J., Neurotoxicity of the psychedelic amphetamine, methylenedioxymethamphetamine, *J. Pharmacol. Exp. Ther.*, 240, 1, 1987.

Schmidt, C. J., Black, C. K. and Taylor, V. L., L-DOPA potentiation of the serotonergic deficits due to a single administration of 3,4-methylenedioxymethamphetamine, p-chloroamphetamine or methamphetamine to rats, *Eur. J. Pharmacol.*, 203, 41, 1991a.

Schmidt, C. J., Black, C. K. and Taylor, V. L., Fadayel, G. M., Humphreys, T. M., Nieduzak, T. R. and Sorensen, S. M., The 5-HT$_2$ receptor antagonist, MDL 28,133A, disrupts the serotonergic-dopaminergic interaction mediating the neurochemical effects of 3,4-methylenedioxymethamphetamine, *Eur. J. Pharmacol.*, 220, 151, 1992.

Schmidt, C. J. and Kehne, J. H., Neurotoxicity of MDMA: Neurochemical effects, in *The Neuropharmacology of Serotonin*, Whitaker-Azmitia, P. M. and Peroutka, S. J., Eds., *Ann. N.Y. Acad. Sci.*, Vol. 600, New York, 1990, pg. 665.

Schmidt, C. J., Levin, J. A. and Lovenberg, W., In vitro and in vivo neurochemical effects of methylenedioxymethamphetamine on striatal monoaminergic systems in the rat brain, *Biochem. Pharmacol.*, 36, 747, 1987.

Schmidt, C. J., Taylor, V. L., Abbate, G. M. and Nieduzak, T. R., 5-HT$_2$ antagonists stereoselectively prevent the neurotoxicity of 3,4-methylenedioxymethamphetamine by blocking the acute stimulation of dopamine synthesis: Reversal by L-DOPA, *J. Pharmacol. Exp. Ther.*, 256, 230, 1991b.

Schultz, W. and Romo, R., Responses of nigrostriatal dopamine neurons in high-intensity somatosensory stimulation in the anesthetized monkey, *J. Neurophysiol.*, 57, 201, 1987.

Sinton, C. M. and Fallon, S. L., Electrophysiological evidence for a functional differentiation between subtypes of the 5-HT$_1$ receptor, *Eur. J. Pharmacol.*, 157, 173, 1988.

Skarsfeldt, T., Electrophysiological profile of the new atypical neuroleptic, sertindole, on midbrain dopamine neurones in rats: Acute and repeated treatment, *Synapse*, 10, 25, 1992.

Skarsfeldt, T. and Larsen, J.-J., SCH 23390 - a selective dopamine D-1 receptor antagonist with putative 5-HT$_1$ receptor agonistic activity, *Eur. J. Pharmacol.*, 148, 389, 1988.

Skarsfeldt, T. and Perregaard, J., Sertindole, a new neuroleptic with extreme selectivity on A10 versus A9 dopamine neurones in the rat, *Eur. J. Pharmacol.*, 182, 613, 1990.

Soghomonian, J.-J., Doucet, G. and Descarries, L., Serotonin innervation in adult rat neostriatum. I. Quantified regional distribution, *Brain Res.*, 425, 85, 1987.

Sorensen, S. M., Humphreys, T. M. and Palfreyman, M. G., Effect of acute and chronic MDL 73,147EF, a 5-HT$_3$ receptor antagonist, on A9 and A10 dopamine neurons, *Eur. J. Pharmacol.*, 163, 115, 1989.

Sorensen, S. M., Humphreys, T. M., Taylor, V. L. and Schmidt, C. J., 5-HT$_2$ receptor antagonists reverse amphetamine-induced slowing of dopaminergic neurons by interfering with stimulated dopamine synthesis, *J. Pharmacol. Exp. Ther.*, 260, 872, 1992.

Sorensen, S. M., Kehne, J. H., Fadayel, G. M. Humphreys, T. M., Ketteler, H. J., Sullivan, C. K., Taylor, V. L. and Schmidt, C. J., Characterization of the 5-HT$_2$ receptor antagonist MDL 100907 as a putative atypical antipsychotic: Behavioral, electrophysiological and neurochemical studies, *J. Pharmacol. Exp. Ther.*, 266, 684, 1993.

Sprouse, J. S. and Aghajanian, G. K., Electrophysiological responses of serotonergic dorsal raphe neurons to 5-HT1A and 5-HT1B agonists, *Synapse*, 1, 3, 1987.

Sprouse, J. S., Bradberry, C. W., Roth, R. H. and Aghajanian, G. K., MDMA (3,4-methylenedioxymethamphetamine) inhibits the firing of dorsal raphe neurons in brain slices via release of serotonin, *Eur. J. Pharmacol.*, 167, 375, 1989.

Sprouse, J. S., Bradberry, C. W., Roth, R. H. and Aghajanian, G. K., 3,4-Methylenedioxymethamphetamine-induced release of serotonin and inhibition of dorsal raphe cell firing: Potentiation by L-tryptophan, *Eur. J. Pharmacol.*, 178, 313, 1990.

Stachowiak, M. K., Bruno, J. P., Snyder, A. M., Stricker, E. M. and Zigmond, M. J., Apparent sprouting of striatal serotonergic terminals after dopamine-depleting brain lesions in neonatal rats, *Brain Res.*, 291, 164, 1984.

Stone, D. M., Johnson, M., Hanson, G. R. and Gibb, J. W., Role of endogenous dopamine in the central serotonergic deficits induced by 3,4-methylenedioxymethamphetamine, *J. Pharmacol. Exp. Ther.*, 247, 79, 1988.

Stone, D. M., Stahl, D. C., Hanson, G. R. and Gibb, J. W., The effects of 3,4-methylenedioxymethamphetamine (MDMA) and 3,4-methylenedioxyamphetamine (MDA) on monoaminergic systems in the rat brain, *Eur. J. Pharmacol.*, 128, 41, 1986.

Strecker, R. E. and Jacobs, B. L., Dopaminergic unit activity during behavior, in *Neurophysiology of Dopaminergic Systems - Current Status and Clinical Perspectives*, Chiodo, L. A. and Freeman, A. S., Eds., Lakeshore Publishing Co., Grosse Pointe, 1987, pg. 165.

Stotz, E. H., Triarhou, L. C., Ghetti, B. and Simon, J. R., Serotonin content is elevated in the dopamine deficient striatum of the weaver mutant mouse, *Brain Res.*, 606, 267, 1993.

Svensson, T. H. and Tung, C.-S., Local cooling of pre-frontal cortex induces pacemaker-like firing of dopamine neurons in rat ventral tegmental area in vivo, *Acta Physiol. Scand.*, 136, 135, 1989.

Svensson, T. H., Tung, C.-S. and Grenhoff, J., The 5-HT$_2$ antagonist ritanserin blocks the effect of pre-frontal cortex inactivation of rat A10 dopamine neurons in vivo, *Acta Physiol. Scand.*, 136, 497, 1989.

Thierry, A. M., Mantz, J., Milla, C. and Glowinski, J., Influence of the mesocortical/prefrontal dopamine neurons on their target cells, in *The Mesocorticolimbic Dopamine System*, Kalivas, P. W. and Nemeroff, C. B., Eds., *Ann. N. Y. Acad. Sci.*, Vol. 537, New York, 1988, pg. 101.

Tork, I., Raphe nuclei and serotonin containing systems, in *The Rat Nervous System, Vol. 2: Hindbrain and Spinal Cord*, Paxinos, G., Ed., Academic Press, New York, 1985, chap. 3.

Tork, I., Anatomy of the Serotonergic System, in *The Neuropharmacology of Serotonin*, Whitaker-Azmitia, P. M. and Peroutka, S. J., Eds., *Ann. N. Y. Acad. Sci.*, Vol. 600, New York, 1990, pg. 9.

Trent, F. and Tepper, J. M., Dorsal raphe stimulation modifies striatal-evoked antidromic invasion of nigral dopaminergic neurons in vivo, *Exp. Brain Res.*, 84, 620, 1991.

Tricklebank, M. D., Interactions between dopamine and 5-HT$_3$ receptors suggest new treatments for psychosis and drug addiction, *TIPS*, 10, 127, 1989.

Tricklebank, M. D., Forler, C. and Fozard, J. R., The involvement of subtypes of the 5-HT$_1$ receptor and of catecholaminergic systems in the behavioural response to 8-hydroxy-2-(di-n-propylamino)tetralin in the rat, *Eur. J. Pharmacol.*, 106, 271, 1985.

Ugedo, L., Grenhoff, J. and Svensson, T. H., Ritanserin, a 5-HT$_2$ receptor antagonist, activates midbrain dopamine neurons by blocking serotonergic inhibition, *Psychopharmacol.*, 98, 45, 1989.

Van der Kooy, D. and Hattori, T., Dorsal raphe cells with collateral projections to the caudate-putamen and substantia nigra: A fluorescent

retrograde double labeling study in the rat, *Brain Res.*, 186, 1, 1980.
Volonte, M., Ceci, A. and Borsini, F., Effect of haloperidol and clozapine on
(+)SKF 10,047-induced dopamine release: Role of 5-HT$_3$ receptors, *Eur. J.
Pharmacol.*, 213, 163, 1992.
Wadenberg, M.-L., Ahlenius, S. and Svensson, T. H., Potency mismatch for
behavioral and biochemical effects by dopamine receptor antagonists:
Implications for the mechanism of action of clozapine, *Psychopharmacol.*, 110,
273, 1993.
Wander, T. J., Nelson, A., Okazaki, H. and Richelson, E., Antagonism by
neuroleptics of serotonin 5-HT$_{1A}$ and 5-HT$_2$ receptors of normal human brain
in vitro, *Eur. J. Pharmacol.*, 143, 279, 1987.
Weinberger, D. R., Berman, K. F. and Chase, T. N., Mesocortical
dopaminergic function and human cognition, in *The Mesocorticolimbic
Dopamine System*, Kalivas, P. W. and Nemeroff, C. B., Eds., *Ann. N. Y. Acad.
Sci.*, Vol. 537, New York, 1988, pg. 330.
Whitaker-Azmitia, P. M. and Peroutka, S. J., Eds., *The Neuropharmacology
of Serotonin, Ann. N.Y. Acad. Sci.*, Vol. 600, New York, 1990.
White, F. J. and Wang, R. Y., Differential effects of classical and atypical
antipsychotic drugs on A9 and A10 dopamine neurons, *Science*, 221, 1054,
1983.
White, F. J. and Wang, R. Y., A10 dopamine neurons: role of autoreceptors
in determining firing rate and sensitivity to dopamine agonists, *Life Sci.*, 34,
1161, 1984.
Wilmot, C. A. and Szczepanik, A. M., Effects of acute and chronic
treatments with clozapine and haloperidol on serotonin (5-HT$_2$) and dopamine
(D$_2$) receptors in the rat brain, *Brain Res.*, 487, 288, 1989.
Wirtshafter, D., Stratford, T. R. and Asin, K. E., Evidence that serotonergic
projections to the substantia nigra in the rat arise in the dorsal, but not the
median, raphe nucleus, *Neurosci. Lett.*, 77, 261, 1987.
Wozniak, K. M., Pert, A. and Linnoila, M., Antagonism of 5-HT$_3$ receptors
attenuates the effects of ethanol on extracellular dopamine, *Eur. J. Pharmacol.*,
187, 287, 1990.
Yamamoto, B. K. and Spanos, L. J., The acute effects of
methylenedioxymethamphetamine on dopamine release in the awake-behaving
rat, *Eur. J. Pharmacol.*, 148, 195, 1988.
Young, K. A., Zavodny, R. and Hicks, P. B., Effects of serotonergic agents
on apomorphine-induced locomotor activity, *Psychopharmacol.*, 110, 97, 1993.
Zhou, F. C., Bledsoe, S. and Murphy, J., Serotonergic sprouting is induced
by dopamine-lesion in substantia nigra of adult rat brain, *Brain Res.*, 556, 108,
1991.

CHAPTER 4

ELECTROPHYSIOLOGICAL AND BIOCHEMICAL INTERACTIONS BETWEEN DOPAMINE AND CHOLECYSTOKININ IN THE BRAIN

Arthur S. Freeman

I. INTRODUCTION

It is generally accepted that altered function of central dopaminergic (DA) systems originating in the midbrain underlies the symptomatology of several neurological and psychiatric disorders. Progressively more detailed understanding of the anatomy of midbrain DA neurons has permitted extraordinary progress in the study of the biochemical and electrophysiological properties, as well as the behavioral significance, of distinct DA pathways. A detailed understanding of the influence of these neurons on brain function will enable specific hypotheses about their roles in normal and pathological conditions to be tested. This may lead, ultimately, to rational and effective treatments of the clinical disorders believed to result from DA system dysfunction.

The majority of rat midbrain DA neurons also contain mRNA for cholecystokinin (CCK). Brain CCK is present mainly in the form of its sulfated carboxyterminal octapeptide (CCK-8S). [Beinfeld, 1981; Dockray, 1980; Rehfeld, 1978] Expression of CCK-8S varies among the different DA projections, where it appears to function as a cotransmitter. Endogenous CCK-8S presumably influences the function of neurons that are also under the influence of DA, but the role of CCK-8S in DA neurotransmission has only begun to be evaluated. In this regard, complex relationships have been reported to occur between the circadian variations in the levels of DA and CCK-8S in forebrain sites innervated by DA neurons. [Schade et al., 1993] The colocalization of CCK-8S with DA further suggests that CCK-8S plays a role in clinical disorders generally associated with DA. The CCK component of DA pathways may, therefore, provide alternative sites of action (*vis-à-vis* DA receptors) at which to manipulate the influence of DA neurons on brain function.

There are two major subtypes of CCK receptors, CCK-A and CCK-B. Studies over the past several years have shed light on the receptor selectivity of the effects of CCK-8S on several measures of DA electrophysiology and neurochemistry, and have set the stage for continued analysis of the role of CCK-8S in DA neurotransmission. CCK-8S affects the release of DA from DA nerve terminals in the nucleus accumbens, caudate nucleus and prefrontal cortex, and also affects the electrophysiological activity of midbrain DA and non-DA neurons, and neurons in

DA-innervated forebrain regions. It is likely, therefore, that CCK-8S plays roles in synaptic and nonsynaptic DA neurotransmission. Thus, most DA neurons are really "DA/CCK-8" neurons, and this dual transmitter status may be essential to their normal function.

II. DA NEURONS CONTAIN CCK-8S

Rat midbrain DA neurons may be classified on the basis of several characteristics and properties: cell body location, axonal projection site, unique electrophysiological properties (e.g., range of firing rates and types of firing patterns), biochemical indices, and pharmacological responsiveness. [Bunney, 1979; Chiodo, 1988; Chiodo et al., 1984] The growing list of instances where monoaminergic neurons contain a peptide(s) includes midbrain DA neurons. It was reported in 1980 that CCK-8 is present within some DA-containing neurons of the rat midbrain. [Hökfelt et al., 1980] This marked DA/CCK-8S coexistence was found in several mesolimbic projections, including the posterior-medial nucleus accumbens, olfactory tubercle, central nucleus of the amygdala, and the bed nucleus of the stria terminalis. Thus, neurochemical lesions of midbrain DA neurons with 6-hydroxydopamine result in a decrease in the levels of not only DA, but also of CCK-8 in these regions. [Gilles et al., 1983; Hökfelt et al., 1980; Marley et al., 1982; Studler et al., 1981; Williams et al., 1981] The presence of CCK-like immunoreactivity in the midbrain is restricted mainly to DA neurons, which suggests that CCK-8 has important regulatory influences on the function of DA neurons and their target sites. [Seroogy et al., 1989a]

Some of these early studies reported a DA/CCK projection to the ventromedial caudate nucleus [Hökfelt et al., 1980; Marley et al., 1982], whereas others did not (see Vanderhaeghen, 1985). With the use of colchicine and more sensitive CCK antisera, it was later discovered that DA/CCK colocalization is actually prevalent in DA cell bodies throughout most of the substantia nigra pars compacta (A9) and throughout the ventral tegmental area (VTA or A10). [Seroogy et al., 1989a] Results from immunohistochemical experiments combined with retrograde tracing led Seroogy et al. to conclude that a large proportion of DA neurons giving rise to the ascending mesotelencephalic projections (i.e., mesostriatal, mesolimbic, mesocortical) contain CCK peptide. [Seroogy et al., 1989a] In accordance with these findings, *in situ* hybridization studies have demonstrated the coexistence of tyrosine hydroxylase mRNA and CCK mRNA in rat mesencephalic neurons. [Savasta et al., 1989; Seroogy et al., 1989b] Despite the presence of CCK-like immunoreactivity in nigrostriatal neurons, this projection does not appear to account for the bulk of CCK in the caudate. [Gilles et al., 1983; Marley et al., 1982] The caudate nucleus also receives a CCK input from the amygdala, but the largest inputs appear to be cortically derived. [Fallon et al., 1983; Meyer et al., 1982, 1988; Morino et al., 1994; Záborszky et al., 1985]

III. CCK RECEPTORS

A. CCK RECEPTOR SUBTYPES

CCK receptors were first classified into the subtypes, CCK-A and CCK-B, based on the relative affinities of CCK fragments for displacement of radiolabelled CCK-33. [Moran et al., 1986] In this nomenclature, "A" refers to alimentary and "B" refers to brain. CCK-A receptors have high selectivity for CCK-8S compared to unsulfated CCK-8 (CCK-8U) and CCK-4. CCK-A receptors are often referred to as peripheral-type CCK receptors because of their prevalence in the pancreas, gallbladder, and vagus nerve. [Innis and Snyder, 1980; for review, see Beinfeld, 1983] CCK-A receptors also occur, however, in the area postrema, nucleus tractus solitarius (partially on vagal afferents), interpeduncular nucleus, central amygdala, and posterior hypothalamus of the rat brain. [Carlberg et al., 1992; Hill et al., 1987, 1988, 1990; Moran et al., 1986, 1990] It was reported recently that CCK-A receptors are also located on DA neurons of the monkey substantia nigra/ventral tegmental area complex and its projection areas. [Graham et al., 1991] In that study, unilateral MPTP-induced degeneration of midbrain DA neurons was associated with a loss of CCK-A receptors in the midbrain and in the ipsilateral nucleus accumbens and medial caudate nucleus. The rat CCK-A receptor has been cloned and appears to have seven membrane-spanning segments, which suggests that it is a G protein-coupled receptor. [Wank et al., 1992a] The regional brain distribution of CCK-A receptor mRNA has been recently evaluated by *in situ* hybridization. Hybridization signals for the CCK-A receptor probe were found to be distributed in several areas, including cortex, olfactory areas, septum, hippocampal areas, hypothalamic nuclei, medial nucleus of the amygdala, interpeduncular nucleus, and dorsal motor nucleus of the vagus. [Honda et al., 1993] This study suggests that, despite the apparent limited distribution of CCK-A receptor expression in brain, many brain areas have the potential to express CCK-A receptors. There may be other brain sites with currently undetectable levels of CCK-A receptor mRNA and receptor synthesis. This would be consistent with results of pharmacological studies discussed below that point to CCK-A receptor-mediated effects in brain areas apparently bereft of CCK-A receptors.

Rat brain and human brain CCK-B receptors have also been cloned and characterized. [Ito et al., 1993; Lee et al., 1993; Wank et al., 1992b] The predicted amino acid sequences for these receptors are reported to be 89% identical. [Ito et al., 1993] Like the rat CCK-A receptor, CCK-B receptors have seven putative transmembrane domains characteristic of G protein-linked receptors. In comparison, CCK-A and CCK-B receptors are 48% identical in their deduced sequences, which differ markedly in the characteristics of their third intracellular loops. [Wank et al., 1992b] CCK-B receptors are widely distributed in the rat brain and do not discriminate greatly among the CCK peptide fragments. [Carlberg et al., 1992; Moran et al., 1986] *In situ* hybridization signals for CCK-B receptor mRNA are strong in many brain regions, which include many of those that also contain

CCK-A receptor mRNA. [Honda et al., 1993] CCK-B receptor mRNA was also prevalent in the caudate nucleus, nucleus accumbens, brain stem sensory nuclei, and throughout the amygdala.

Although almost all neuronal CCK is present as CCK-8S (a mixed A/B agonist), the comparison of CCK-8S effects with those of CCK-8U and CCK-4 in a given test can provide preliminary evidence of a role for CCK-A or CCK-B receptors. Thus, the high affinity of CCK-8U and CCK-4 for CCK-B receptors compared to CCK-A receptors has been exploited to determine the receptor-selectivity of various CCK-8S effects. In addition, the development of selective receptor antagonists has made it possible to assess more directly the involvement of specific receptor subtypes in the effects of CCK-8S.

B. SELECTIVE CCK RECEPTOR ANTAGONISTS

Until recently, the only CCK antagonists available (e.g., proglumide, benzotript) were of low potency and were nonselective with regard to CCK receptor subtype. A number of selective antagonists have now been developed by pharmaceutical companies. Selective antagonists for the CCK-A receptor include lorglumide (CR 1409; Rotta) and L-364,718 (MK-329, devazepide; Merck). L-365,260 and L-740,093 (Merck), PD134308 (CI-988) and PD135158 (Parke-Davis), and LY262691 and related compounds (Lilly) are selective CCK-B receptor antagonists. These compounds have been used in the identification of receptor subtypes that mediate some of the biological effects of CCK-8S, and in evaluation of possible therapeutic applications of CCK antagonists. [Wettstein et al., 1994]

IV. ELECTROPHYSIOLOGICAL EFFECTS OF CCK PEPTIDES AND ANTAGONISTS

A. EFFECTS OF CCK PEPTIDES AND CCK ANTAGONISTS ON DA NEURONAL ACTIVITY

1. CCK Peptides

Shortly after the discovery of DA/CCK-8 coexistence, short-lived (several minutes) changes in the firing rate and firing pattern of rat midbrain DA neurons were observed after i.v. administration of CCK-8S and CCK-7S (whose biological activity is similar to CCK-8S). [Chiodo et al., 1987; Freeman and Bunney, 1987; Freeman and Chiodo, 1987, 1988; Hommer et al., 1985, 1986; Kelland et al., 1991; Skirboll et al., 1981] In the first study, it was observed that CCK-7S produced an immediate increase in the firing rate of a majority of DA neurons sampled in a region of extensive DA/CCK-8S coexistence (medial A9). [Skirboll et al., 1981] This was consistent with the anatomical localization of CCK-like immunoreactivity to mesolimbic DA neurons, many of which originate in medial A9. These increases in firing rate were often associated with dramatic increases in burst-firing. CCK-7U had no effects on DA cell firing.

It was later shown with the method of antidromic activation that a subpopulation of DA neurons in a putatively CCK-poor (nigrostriatal) projection were also excited by CCK-8S (but not CCK-8U). [Freeman and Chiodo, 1988; Freeman et al., 1988] At the time, this suggested that cell responsiveness to CCK-8S was not necessarily conditional upon DA/CCK-8S coexistence. Whole cell patch-clamp experiments on acutely dissociated nigral DA neurons demonstrated that CCK-8S (but not CCK-4) caused a depolarization by induction of an inward cationic current. [Wu and Wang, 1994] Because most DA neurons are now thought to contain CCK mRNA and variable amounts of CCK-like immunoreactivity [Seroogy et al., 1989a,b], these studies suggest that DA cell responsiveness to CCK-8S cannot be predicted on the basis of the presumed presence or absence of DA/CCK-8S coexistence as once believed.

In A10, DA neurons underwent increases, decreases, or no change in firing rate in response to i.v. CCK-7S or CCK-8S. [Freeman and Bunney, 1987; Hommer et al., 1986; Kelland et al., 1991; Skirboll et al., 1981] CCK-7U, CCK-8U and CCK-4 had no effect on firing rate in these studies. The requirement for sulfation for the effects of CCK peptides on DA cell firing rate suggests that these effects are mediated via occupation of CCK-A receptors (see CCK Antagonists, below).

To control for possible peripheral or remote sites of action of i.v. CCK-8S, the peptide was applied by microiontophoresis. This local application resulted in stimulation of the firing rate and the burst-firing of DA neurons both in medial A9 and in A10 [Skirboll et al., 1981], and of a subpopulation of identified nigrostriatal DA neurons. [Freeman et al., 1988] No inhibitory effects were observed following this route of administration, which suggests that the frequent inhibitory effects of i.v. CCK-7S and CCK-8S on A10 DA cells are mediated indirectly. In *in vitro* slices of rat midbrain, CCK-8S again was observed to stimulate the firing rate of medial A9 and A10 DA neurons. [Brodie and Dunwiddie, 1987; Freeman et al., 1989; Stittsworth and Mueller, 1990] This effect, however, was not accompanied by the increase in burst-firing observed *in vivo*. [Stittsworth and Mueller, 1990]

The lack of inhibitory effects of CCK-7S and CCK-8S microiontophoresis on the firing rate of any tested A10 DA cell suggests that i.v. CCK-induced inhibition is exerted at a location remote from the affected DA cell. Whether administered i.v. or microiontophoretically, however, excitatory responses include increases in firing rate and increases in burst-firing. Despite these similarities, it was questioned how amounts of peptide sufficient to cause these effects could cross the blood-brain barrier after such very small i.v. doses (1-20 μg/kg). Because vagotomy blocks certain behavioral effects of CCK-8S, Hommer and colleagues evaluated the roles of the vagus nerve and its central relay site, the nucleus tractus solitarius (NTS), in the excitatory effects of i.v. CCK-8S on rat medial A9 DA neurons. CCK-8S-induced excitations were not reduced by vagotomy, but were reduced significantly in rats with lesions of efferents of the nucleus tractus solitarius. [Hommer et al., 1985] These results point to direct effects of i.v. CCK-8S in the brain to stimulate

DA cell firing rate; one site is likely the DA cell itself, whereas another may be on NTS efferents.

2. Effects of Antagonists Alone and Against CCK-8S

The CCK antagonist proglumide blocks the excitatory effects of CCK-8S on medial A9 and A10 DA cell firing rate. [Chiodo and Bunney, 1983; Chiodo et al., 1987; Freeman and Bunney, 1987] Proglumide, however, does not distinguish between CCK-A and CCK-B receptors. The excitatory effects of CCK-8S on mesoaccumbal DA cell firing rate were later found to be blocked by pretreatment with the CCK-A receptor antagonist, devazepide, but not the CCK-B receptor antagonist, L-365,260. [Kelland et al., 1991] Similarly, the CCK-8S-induced inward current in acutely dissociated nigral DA neurons was inhibited by lorglumide but not by PD135158. [Wu and Wang, 1994] These results are consistent with the results from several studies referred to above where sulfation was required for the excitatory effects of CCK-7S and CCK-8S. In the study by Kelland et al. of mesoaccumbal DA neurons, i.v. CCK-8S also inhibited a subpopulation of these neurons. Pretreatment with L-365,260, but not devazepide, significantly reduced the number of mesoaccumbal DA cells inhibited by CCK-8S. [Kelland et al., 1991] Although this would appear to implicate a role for CCK-B receptors in the observed inhibitory effects of CCK-8S, the lack of effect of CCK-8U and CCK-4 on these neurons argues against this interpretation. [Kelland et al., 1991] It is therefore not clear what receptor mediates this response. Though speculative, perhaps costimulation of CCK-A and CCK-B receptors is required for certain effects of CCK-8S.

To determine if endogenous CCK-8S exerts tonic effects on DA cell firing rate, the effects of proglumide and benzotript, and later, selective CCK-A and CCK-B receptor antagonists were evaluated. Most of these studies failed to identify acute effects of CCK antagonists alone on DA cell electrophysiology in untreated rats. Proglumide and benzotript had no significant effects on firing rate. [Chiodo and Bunney, 1983, 1987; Chiodo et al., 1987; Freeman et al., 1988; Kelland et al., 1991] Neither devazepide, lorglumide, L-365,260, nor PD 134308 had significant effects on firing rate. [Kelland et al., 1991; Meltzer et al., 1993; Zhang et al., 1991]

In contrast to the electrophysiological results summarized above, profound acute effects have been reported for the diphenylpyrazolidinone CCK-B antagonist LY262691 on the electrophysiological activity of rat midbrain DA neurons. Two hours after an injection of LY262691 (3-30 mg/kg, i.p.), the average firing rates of DA cells in the A9 and A10 populations were significantly decreased. [Rasmussen et al., 1991] The firing rates of individual A9 and A10 DA neurons monitored continuously before and after drug administration were also observed to decrease after LY262691 (30 mg/kg, i.p.). [Rasmussen et al., 1993a] Meltzer et al., however, did not observe consistent effects of PD 134308 (0.1-6.4 mg/kg, i.v.) on A9 and A10 DA cell firing rates. [Meltzer et al., 1993] The reason for the discrepancy between the reported effects of LY262691 and PD 134038 on DA cell firing rate is not clear, but is due presumably to uncharacterized differences in the properties of these CCK-B antagonists. It remains to be demonstrated that CCK-B antagonists other than those of the diphenylpyrazolidinone class can reduce the firing rate of midbrain DA neurons.

3. DA Cell Population Studies

The cells-per-track population sampling method has been applied to the study of CCK and CCK antagonist effects on DA neurons. In this procedure, a recording electrode is passed repeatedly through a stereotaxically defined region of the rat midbrain, and the numbers of spontaneously active A9 and A10 DA neurons are counted. The results are expressed as the number of active cells per electrode track. Acute i.v. administration of CCK-8S (8 μg/kg, i.v.), but not CCK-8U (8 or 20 μg/kg), produced a twofold increase in the number of spontaneously active A9 and A10 DA neurons compared to saline controls. [Freeman and Chiodo, 1988] Acute administration of the nonselective CCK antagonist, proglumide, did not alter the number of spontaneously active A9 or A10 DA neurons. [Chiodo and Bunney, 1987] Similar to proglumide, acute administration of the selective CCK-A antagonists, lorglumide and LY 219057 did not affect the number of active DA neurons. [Rasmussen et al., 1993a; Zhang et al., 1991]

In contrast to these results, Rasmussen and colleagues reported significant effects of selective CCK-B antagonists on the number of spontaneously active DA neurons. A single injection of each of four selective diphenylpyrazolidinone CCK-B antagonists, LY262691, LY262684, LY191009 and LY242040, was reported to decrease the number of spontaneously active A9 and A10 DA neurons 2 hr later. [Rasmussen et al., 1991, 1993a,b] As noted above, LY262691 was also shown to inhibit the firing rates of individual A9 and A10 DA neurons. [Rasmussen et al., 1993a] This latter finding supported the previous conclusion that the effects of LY262691 on DA cell population activity are not due to depolarization inactivation of the DA cell membrane. [Rasmussen et al., 1991] In this study, it was also reported that L-365,260 (3 and 10 mg/kg, i.p.) similarly reduced the number of spontaneously active A10 DA neurons (A9 not reported). These investigators found subsequently that direct injection of LY262691 into the caudate nucleus or nucleus accumbens reduced the number of spontaneously active A9 and A10 DA neurons, respectively. [Rasmussen et al., 1993b] Conversely, lesions of these brain regions had corresponding blocking effects on the ability of i.p. LY262691 to reduce the number of active DA neurons in A9 and A10. This study suggests that the caudate nucleus and nucleus accumbens are the major sites of action for the effects of systemically administered LY262691 on DA cell population activity. Taken together, these studies of LY262691 and related compounds suggest that CCK-8S, through interactions at forebrain CCK-B receptors, exerts a tonic stimulatory influence on the number of spontaneously active midbrain DA neurons. Two quinazolinone CCK-B antagonists (LY 202769, LY247348) were also reported to decrease the number of spontaneously active A9 and A10 DA neurons in anesthetized rats two hours after i.p. administration (Rasmussen et al., 1994).

Reduction in the number of spontaneously active rat A10 DA neurons is associated with clinical efficacy of antipsychotic drugs, whereas a similar effect in A9 is associated with liability for extrapyramidal side-effects. Although LY262691 reduced the number of active cells in A9 and A10, this compound was reported to be devoid of cataleptogenic effects in rats, which suggests low liability for extrapyramidal side-effects. [Rasmussen et al., 1993a] Perhaps manipulation of CCK-B receptors will prove to be a useful approach to the development of novel antipsychotic drugs.

Acute i.v. injection of either proglumide, lorglumide or devazepide reversed the reduction in the numbers of spontaneously active A9 and A10 DA neurons produced by repeated haloperidol administration. [Chiodo and Bunney, 1987; Jiang et al., 1988; Minabe et al., 1991] This implicates a role for endogenous CCK-8S in the effects of repeated haloperidol on DA cell population activity. Acute microinjection of lorglumide into the nucleus accumbens, but not the caudate nucleus, also reversed the effects of repeated haloperidol on A9 and A10 DA neurons, which implicates the nucleus accumbens as a critical site for this influence of endogenous CCK-8S.

Two studies evaluated the effects of *repeated* treatment with CCK antagonists on the number of spontaneously active DA neurons in rats. Proglumide (21 days, 1-10 mg/kg, p.o.) increased the number of spontaneously active A9 and A10 DA cells. [Chiodo and Bunney, 1987] Lorglumide (14 days, 0.5-5 mg/kg, i.p.) increased the number of active A10 DA cells, whereas small nonsignificant increases were observed in the A9 area. [Zhang et al., 1991]

B.   CCK MODULATION OF IMPULSE-REGULATING D2 DA AUTORECEPTORS

1. CCK Peptides

Stimulation of impulse-regulating D2 DA autoreceptors results in inhibition of DA cell firing rate. In addition to its short-lived effects on firing rate, CCK-8S can alter the potency of direct-acting D2 DA receptor agonists to inhibit DA neuronal firing rate. [Brodie and Dunwiddie, 1987; Chiodo et al., 1987; Hommer and Skirboll, 1983; Hommer et al., 1986; Stittsworth and Mueller, 1990] Some i.v. studies suggest that this effect of CCK-8S is restricted to the sulfated form of the peptide, and implicate a role for CCK-A receptors in this modulatory effect of CCK-8S. [Freeman and Bunney, 1987; Freeman and Chiodo, 1988; Kelland et al., 1991] There is not unanimity on this point, however, as another study reported that i.v. CCK-8U and CCK-4 also increased the inhibitory effects of the DA agonist apomorphine. [Hommer et al., 1986] This latter finding suggested, therefore, that CCK-B receptors mediate the modulatory effect of CCK peptides on DA autoreceptor sensitivity. The modulatory effects of CCK-8S on D2 receptor-mediated inhibition of DA neurons are more enduring than the transient effects of

CCK-8S on firing rate described above. [Brodie and Dunwiddie, 1987; Chiodo et al., 1987; Freeman and Chiodo, 1988; Hommer and Skirboll, 1983; Hommer et al., 1986; Stittsworth and Mueller, 1990] At this time, the second messenger systems responsible for these prolonged effects are not known.

2. CCK Antagonists

Studies with CCK antagonists have also not led to a clear consensus as to the CCK receptor that modulates DA cell sensitivity to D2 DA agonists. For example, acute i.v. administration of a low dose (0.1 mg/kg, i.v.) of the CCK-A antagonist devazepide, but not the CCK-B antagonist L-365,260, blocked the ability of CCK-8S to enhance the potency of the D2/D3 agonist quinpirole to inhibit firing rate. [Kelland et al., 1991] Studies of the effects of antagonists *alone*, however, have yielded contradictory results. A low dose (0.1 mg/kg, i.v.) of either devazepide or L-365,260 did not alter the quinpirole dose-response curve for inhibition of the firing rates of identified mesoaccumbal DA cells. [Kelland et al., 1991] In a similar study, devazepide (0.6 and 4.16 mg/kg) did not alter the dose-response curve for apomorphine-induced inhibition of A9 DA neurons, whereas PD 134308 (0.6 and 6.4 mg/kg, i.v.) significantly reduced the potency of apomorphine. [Meltzer et al., 1993]

C. EFFECTS OF CCK PEPTIDES ON MIDBRAIN NON-DA NEURONS

In the substantia nigra pars reticulata (SNr), non-DA neurons (interneurons and/or collaterals of output neurons) exert an inhibitory influence on the activity of DA neurons. [Anderson et al., 1993; Grace and Bunney, 1979] Non-DA interneurons in the A10 area appear to influence A10 DA neuronal activity in a similar manner. [Waszczak and Walters, 1980] In addition, SNr neurons are excited by somatodendritically released DA. [Waszczak and Walters, 1986] It was therefore considered likely that non-DA SNr and A10 neurons are also influenced by CCK-8 released somatodendritically from DA/CCK-8 neurons (see Section V below). In this way, CCK-8, like DA, may indirectly influence the activity of DA neurons via interactions with non-DA mesencephalic neurons. The SNr is the source of non-DA nigrothalamic and nigrotectal neurons. In addition to sending axon collaterals back to the substantia nigra, these basal ganglia output neurons are important links in the extrapyramidal motor system. [Deniau et al., 1982] The electrophysiological activities of these neurons (identified by antidromic stimulation) are modulated by DA. [Waszczak et al., 1984] Thus, the possibility also exists for "local hormone" effects of CCK-8 on the activity of these output neurons of the SNr.

Initial study of the effects of CCK peptides on non-DA neurons in the substantia nigra pars reticulata has indeed revealed effects on firing rate. Both i.v. (8 μg/kg, i.v.) and local microiontophoretic administration of CCK-8S inhibited the firing rates of about half the cells tested. [Freeman and Zhang, 1993; Zhang and Freeman, 1994] The time-course and magnitude of these inhibitions after i.v. administration were similar to those observed for i.v. CCK-8S-induced excitation of midbrain DA neurons. Although speculative, this suggests that the effect of i.v. CCK-8S on non-DA nigral neurons contributes, via disinhibition, to its excitatory effects on DA neurons. In contrast, i.v. CCK-4 and CCK-8U (8-64 µg/kg, i.v.) produced a gradual and persistent (>10 min) stimulation of the firing rate of non-DA neurons. Because the i.v. administration of these unsulfated peptides does not exert significant effects on DA cell firing rate, their effects on non-DA neurons may lead to unrecognized modulatory effects on DA cells. Microiontophoresis of CCK-4 and CCK-8U never excited the firing rates of non-DA cells, but inhibited the activity of about one-third of the sampled cells. These results suggest that the stimulatory effects of i.v. CCK-4 and CCK-8U on non-DA neuronal activity are mediated indirectly, but that CCK-B receptor stimulation may play a role in the effects of CCK-8S released locally in the midbrain.

D.   ELECTROPHYSIOLOGICAL EFFECTS OF CCK-8S IN THE FOREBRAIN

Microiontophoretic application of CCK-8S stimulates the firing rates of neurons in DA-innervated regions, including the prefrontal cortex, nucleus accumbens, amygdala, and caudate nucleus (although a relatively low percentage of caudate neurons were activated). [Chiodo and Bunney, 1983; DeBonnel and de Montigny, 1988; Rouillard et al., 1989; Wang and Hu, 1986; Wang et al., 1988; White and Wang, 1984; Yim and Mogenson, 1991] In contrast, DA is generally inhibitory to these neurons. Yim and Mogenson evaluated the contributions of DA and CCK-8S to the effects of DA/CCK-8S neuronal stimulation on postsynaptic responses. [Yim and Mogenson, 1991] In their elegant study, they showed that brief electrical stimulation of A10 (10 pulses at 10 Hz) produces effects on accumbal unit activity due to the combined opposing effects of released CCK-8S and DA. Microiontophoretic application of CCK-8S and DA produced similar effects. Initial work by Liang et al. suggests that CCK-8S modulates D2-mediated, but not D1-mediated, electrophysiological effects in the nucleus accumbens. [Liang et al., 1991]

V. BIOCHEMICAL INTERACTIONS BETWEEN DA AND CCK

Biochemical studies have also evaluated the importance of CCK receptors for modulation of DA cell function in the basal ganglia. *In vivo* push-pull cannula perfusion, *in vivo* microdialysis and *in vivo* electrochemical methods have been used, as have *in vitro* methods. Many of these studies are summarized below. The results obtained from the variety of methods and experimental conditions used cannot yet be clearly integrated into a common model. Discrepancies between these studies are due presumably to inherent differences in the methodologies used, anatomical considerations (including the presence of colocalized DA and CCK-8S or separate inputs, and the presence of CCK receptor subtypes in the specific site sampled in a given study), route of drug administration, and the influence of phasic vs. tonic neuronal activity on the effects of receptor stimulation (i.e., the "state" of the tissue"). Nevertheless, the results do show modulatory effects of CCK-8S on indices of DA release, especially in the nucleus accumbens. At the molecular level, CCK-8S has been shown to act indirectly to eliminate cAMP-dependent phosphorylation of a DA- and cAMP-regulated phosphoprotein (DARPP-32) in striatal neurons. [Snyder et al., 1993] Future research will likely focus on interactions between the transduction mechanisms utilized by DA and CCK-8S, and the impact of such interactions on cell function.

A. EFFECTS OF CCK RECEPTOR STIMULATION ON DA RELEASE IN THE NUCLEUS ACCUMBENS

Several studies suggest that potassium-evoked release of DA in the nucleus accumbens is enhanced by CCK-A, but not by CCK-B, receptor stimulation. [Gerhardt et al., 1989; Marshall et al., 1990, 1991; Vickroy et al., 1988] The *in vitro* study of Marshall et al. showed that, in contrast to CCK-A receptor-mediated stimulation of DA release from the posterior nucleus accumbens (where DA and CCK-8S are colocalized), CCK-8S inhibited DA release in the anterior nucleus accumbens (where DA and CCK-8S are not colocalized) due to CCK-B receptor stimulation. [Marshall et al., 1991] This study, as well as others cited below, demonstrates differences in DA/CCK interactions between the anterior and posterior nucleus accumbens. These differences stem, presumably, from the organization of DA and CCK-8S inputs to these areas.

An *in vivo* microdialysis study in anesthetized rats showed that CCK-8S, but not the CCK-B receptor agonist CCK-4 (each administered via the microdialysis probe), increased the *spontaneous* overflow of DA in the dialysate from the anterior but not the posterior nucleus accumbens. [Ruggeri et al., 1987] This effect was attributed to stimulation of "peripheral type" (i.e., CCK-A) CCK receptors. Different conclusions were reached in an *in vivo* microdialysis study in freely moving rats. The CCK-B agonist BC264 (applied via the microdialysis probe) decreased the basal overflow of DA from the anterior nucleus accumbens of freely moving rats. [Ladurelle et al., 1993] In this study, CCK-8S increased DA overflow in the posterior nucleus accumbens.

Basal overflow of DA into the extracellular space of the posterior nucleus accumbens (measured by *in vivo* electrochemistry) has been reported to be unaffected by administration of CCK-8S into this site. [Gerhardt et al., 1989] Another *in vivo* electrochemical study found that basal DA overflow in the medial nucleus accumbens was inhibited by i.v. administration of CCK-8S. [Lane et al., 1986] In an *in vivo* microdialysis study, Laitinen et al. found that the only effect of intra-A10 injection of CCK-8S was to increase the overflow of DA metabolites in the posterior nucleus accumbens after the highest dose tested (10 nmoles). [Laitinen et al., 1990] In contrast, in an *in vivo* microdialysis study in freely moving rats, injection of CCK-8S (1-100 pmoles) but not CCK-4 into the VTA resulted in increases in DA overflow in the posterior nucleus accumbens and amygdala. [Hamilton and Freeman, 1994, 1995] Another microdialysis study in freely moving rats revealed that intraperitoneal injection of CCK-8S, but not CCK-4, increases dopamine turnover in the posterior accumbens. [Kariya et al., 1994] These latter two studies suggest that stimulation of CCK-A receptors enhances the activity of mesoaccumbal DA neurons. This is consistent with the conclusion that CCK-A receptors mediate the direct stimulatory effects of CCK-8S on the firing rate of VTA DA neurons (see Section IV A.1.). In other studies (*in vitro* slice, and *in vivo* push-pull perfusion), CCK-8S but not CCK-8U was found to stimulate basal DA release, but to inhibit potassium-stimulated release of DA, in the medial or posterior nucleus accumbens. [Voigt and Wang, 1984; Voigt et al., 1985, 1986]

B. EFFECTS OF CCK RECEPTOR STIMULATION ON DA RELEASE IN THE CAUDATE NUCLEUS

Some studies have evaluated the effects of CCK-8S on DA release in the caudate nucleus. CCK-8S enhanced the stimulatory effect of potassium on the release of tritium from slices of rat caudate preloaded with [^{3}H]DA but did not alter the spontaneous efflux of tritium. [Starr, 1982] Marshall and coworkers found no effect of CCK-8S on the potassium-stimulated release of DA from slices of rat caudate nucleus. [Marshall et al., 1990] CCK-8S, but not CCK-8U, inhibited the basal and electrically stimulated tritium overflow from slices of cat caudate preloaded with [^{3}H]DA, which suggests mediation by CCK-A receptors. [Markstein and Hökfelt, 1984] Another group reported CCK-B receptor-mediated inhibitory effects of s.c. or intracerebroventricular administration of CCK-8S on postmorten accumbal levels of 3-methoxytyramine, an index of DA release. [Altar and Boyar, 1989; Altar et al., 1988]

An *in vivo* electrochemical study in rats found that CCK-8S but not CCK-8U enhanced the potassium-stimulated overflow of DA into the extracellular space, whereas no effect was observed on basal DA overflow. [Gerhardt et al., 1989] The effects of CCK-related peptides on DA overflow in the caudate have also been evaluated with *in vivo* microdialysis in freely moving rats. Neither CCK-8S nor CCK-4 had any effect on the overflow of DA or two of its metabolites, DOPAC and HVA. [Kariya et al., 1994] In another study, it was concluded that the CCK-8S analog, ceruletide, inhibited the depolarization-induced overflow of DA by an action at central CCK-B receptors. [Kihara et al., 1992] In this regard, it has been suggested that inhibitory effects of CCK-8S on DA release may be due to a depolarization block of DA neurons. [Kihara et al., 1992; Lane et al., 1986] In an *in vivo* push-pull perfusion experiment in cats, the nigral application of CCK-8S resulted in an increase in the release of newly synthesized [^{3}H]DA (from perfused [^{3}H]tyrosine) in the caudate nucleus. [Artaud et al., 1989]

C. EFFECTS OF DA RECEPTOR STIMULATION ON THE RELEASE OF CCK-8S

Microdialysis studies in rats have demonstrated calcium-dependent depolarization-induced increases in the overflow of CCK-like immunoreactivity (CCK-LI) in the nucleus accumbens and frontal cortex. [Hurd et al., 1992; Takita et al., 1989] Overflow of CCK-LI in the caudate nucleus is also increased by depolarization. [Butcher et al., 1989] In contrast, amphetamine produced only a small increase in basal CCK overflow in the nucleus accumbens [Hurd et al., 1992]. These results are consistent with the report by Emson et al. [1980] that rat brain CCK-8 is enriched in vesicular fractions and released in a calcium-dependent manner.

The release of CCK-8S in DA terminal areas is influenced by DA agonists. For example, in slices of rat nucleus accumbens, stimulation of D2/D3 receptors with quinpirole reduces the potassium-stimulated release of CCK-LI. [Martin et al., 1986] Release of endogenous DA with amphetamine had no effect on basal release of CCK-LI but suppressed the depolarization-evoked release of CCK-LI from striatal slices; this effect was attributed to D2 receptor stimulation. [Hutchison et al., 1986] Local application (via *in vivo* push-pull cannulae) of quinpirole also reduces the release of CCK-LI in the cat caudate nucleus. [Artaud et al., 1989] In the latter study, DA and the D1 agonist SKF 38393 stimulated the release of CCK-LI. Earlier studies from one group suggested that D1 receptor stimulation inhibits, whereas D2 receptor stimulation enhances the release of CCK-LI from caudate slices. [Conzelman et al., 1984; Meyer and Krauss, 1983; Meyer et al., 1984] In a recent report, however, Brog and Beinfeld described a stimulatory effect of D1 receptor stimulation on the *in vitro* release of CCK-LI from the caudate nucleus. [Brog and Beinfeld, 1992] These authors attributed the discrepancy with previous studies to differences in experimental protocol and in the concentrations of DA agonists used. In addition, they suggested that the stimulatory effects of DA agonists on the release of CCK-LI are mediated not by D1 receptors linked to adenylate cyclase, but by lower affinity D1 receptors linked to inositol phosphate metabolism.

Similar to DA, CCK-8 can be released somatodendritically from rat midbrain slices. The bulk of this released CCK corresponds to CCK-8S (>93% of total CCK-immunoreactivity). [Freeman et al., 1991] Basal release of CCK-8S was very low in this *in vitro* study, but was stimulated 3-fold by potassium (30 mM)-stimulation. Upon exposure to this concentration of potassium, DA cells in midbrain slices cease to fire spontaneous action potentials almost immediately, presumably due to induction of depolarization block. [Freeman and Chiodo, unpublished observation] Stimulation of inhibitory impulse-regulating somatodendritic DA autoreceptors with the D2/D3 DA agonist, quinpirole (1-100 μM), partially reduced this potassium-stimulated CCK-8S release, whereas stimulated release was abolished in calcium-free medium. A conclusion of this study was that depolarization-induced release of CCK-8S by 30 mM potassium is not impulse-dependent, but is calcium-dependent and is modulated by DA autoreceptors.   Although the operation of a corelease mechanism for somatodendritic DA and CCK-8S is unproven, this finding is consistent with evidence that somatodendritic release of DA is also less dependent on impulse flow than is terminal release. [Cheramy et al., 1981; Kalivas and Duffy, 1991; Nissbrandt et al., 1989] Marked dependency on impulse flow, however, has also been observed. [Santiago and Westerink, 1991; Tanaka et al., 1992] With regard to the former possibility, the mechanism for depolarization-stimulated release of dendritic DA has been attributed to activation of dendritic calcium conductances. [Llinas et al., 1984]

CCK-LI is released from cat substantia nigra *in vivo* (determined with the push-pull perfusion method). [Artaud et al., 1989] Local application of tetrodotoxin reduced the spontaneous release of nigral CCK-LI, but not to the extent that caudate nucleus CCK-LI release was reduced by intracaudatal tetrodotoxin. Similar to the studies on dendritic DA release, these results imply the existence of impulse-dependent and impulse-independent components of nigral release of CCK-LI in the cat. These investigators also obtained evidence for corelease of CCK-LI and DA in the caudate nucleus in response to activation of DA cell firing rate.

Because CCK-8S coexists with DA and directly affects DA neuronal electrophysiology (see Section IV) and release of somatodendritic DA [Artaud et al., 1989], it appears appropriate to conceptualize direct CCK-8S-mediated effects on DA neurons as mediated by CCK autoreceptors. These findings are also relevant to the possibility that, via nonsynaptic somatodendritic release, CCK-8S may, like endogenous DA, influence the activity of neighboring non-DA neurons. As discussed above, many of these non-DA neurons are inhibitory to DA neurons, and are influenced by i.v. and local administration of CCK peptides.

VI. BEHAVIORAL INTERACTIONS BETWEEN DA AND CCK

CCK-8S has been shown to have effects on behaviors mediated by DA in the nucleus accumbens.   As noted above, this is a major site of DA/CCK-8S colocalization. The accumbens is intimately involved in the reinforcing properties of drugs and of brain stimulation, and is a mediator of the locomotor-stimulatory effects of DA agonists. These considerations have implications for the functional roles that endogenous CCK-8S may play as a consequence of its electrophysiological and biochemical effects on the mesoaccumbal DA pathway. CCK-8S and DA have been shown by several behavioral measures to interact in complex ways in the nucleus accumbens. [Vaccarino, 1994]

A. REINFORCING PROPERTIES
The conditioned place preference paradigm is one of several tests used to determine if a drug has reinforcing properties. In this test, the time spent in an environment associated with effects of a drug is compared with time spent in an

environment associated with the drug vehicle. Thus, it is assessed if the stimulus properties of a drug become associated with the environment in which prior drug administration took place. Infusion of CCK-8S into the VTA does not produce a conditioned place preference in rats but it potentiates the conditioned place preference induced by s.c. amphetamine. [Pettit and Mueller, 1989] This suggests that CCK-8S has neuromodulatory effects in the VTA to influence the reinforcing effects of amphetamine.

In the intracranial self-stimulation method (ICSS), rats perform a response to self-administer electrical stimulation to a brain region into which an electrode is implanted. Because the mesolimbic DA system is an important "reward pathway," ICSS is often measured after electrode implantation into the medial forebrain bundle (MFB) or VTA. The effect of drug administration on ICSS gives an indication of the reinforcing properties of the drug. Injection of CCK-8S into the posterior nucleus accumbens resulted in an increase in MFB ICSS, whereas injection into the anterior accumbens or into a lateral ventricle decreased ICSS. [De Witte et al., 1987] These results suggest that CCK-8S enhances the rewarding effects of ICSS in the posterior accumbens, where DA and CCK-8S are colocalized, but that other actions after intraventricular injection prevent the reinforcing action. In accordance with the effects of CCK-8S in the posterior accumbens, ICSS through electrodes in the VTA was attenuated by injection of proglumide into the posterior accumbens. [Vaccarino and Vaccarino, 1989] In contrast to administration into the posterior accumbens, microinjection of CCK-8S into the VTA decreased ICSS. This effect was hypothesized to result from a decrease in DA neurotransmission, possibly due to inhibitory effects of CCK-8S on the firing rate of a subset of A10 DA neurons or to enhanced function of inhibitory impulse-regulating DA autoreceptors. [Rompré and Boye, 1993]

The discriminative stimulus effects of cocaine in rats were attenuated by systemic administration of devazepide, but not by CI-988, a finding that suggested that CCK-A antagonists should be studied as potential antagonists of the subjective effects of cocaine in humans. [Massey et al., 1994] Although the discriminative stimulus and ICSS paradigms measure different stimulus properties of drugs, the inhibitory effects of devazepide on the discriminative stimulus effects of cocaine may be related mechanistically with the facilitatory effects of CCK-8S on ICSS described above for injection into the posterior accumbens. For example, DA and CCK-8S are colocalized in the posterior accumbens, a site where CCK-A receptor stimulation enhances the evoked release of DA (see Section V A.). The site of action for the effects of devazepide on cocaine discrimination, however, is not known. In contrast to its effects in rats, devazepide did not attenuate the discriminative stimulus effects of cocaine in squirrel monkeys. [Spealman, 1992] This may represent species differences in drug effects (which may not bode well for the "cocaine antagonist" hypothesis noted above). Further work is needed to clarify this interesting issue.

B. LOCOMOTOR BEHAVIOR

Similar to the study of the effects of CCK-8S on DA biochemistry, the study of the effects of CCK-8S on DA-mediated locomotor behavior has benefited from a subregional analysis in the nucleus accumbens and an understanding of the chemical neuroanatomy of this nucleus. CCK-8S microinjected into the rat accumbens has been reported to either attenuate or potentiate the stimulatory effects of DA agonists on locomotor behavior. [Crawley et al., 1985b; Vaccarino and Rankin, 1989; Van Ree et al., 1983; Weiss et al., 1988] The effect of CCK-8S on similarly applied DA, however, was shown to be related to the site of injection in the accumbens. Injection at sites shown previously to have extensive DA/CCK-8S colocalization resulted in potentiation of DA-induced hyperlocomotion. These areas correspond to the posterior region of the accumbens. Lack of potentiation or attenuation was observed at anterior sites. [Crawley et al., 1985a] The stimulatory effects of s.c. amphetamine on locomotor activity were also attenuated by microinjection of CCK-8S into the anterior accumbens, whereas posterior injection resulted in potentiation of the response to amphetamine. [Vaccarino and Rankin, 1989]

Crawley investigated the receptor selectivity of the potentiating effects of CCK-8S on DA-induced hyperlocomotion in rats. [Crawley, 1992] Microinjection of the CCK-A receptor antagonist devazepide into the posterior accumbens blocked the potentiation by CCK-8S of DA-induced hyperlocomotion, whereas the CCK-B receptor antagonists, CI-988 and L-365,260, did not significantly antagonize this effect of CCK-8S. This is consistent with the lack of effect in this area of the CCK-B receptor agonist, CCK-8U. [Crawley et al., 1986]. (Hypolocomotion in mice following i.p. injection of CCK-8S was also reversed by devazepide but not L-365,260. [O'Neill et al., 1991] This effect of systemically administered CCK-8S, however, appears to be mediated by peripheral CCK receptors. [Crawley et al., 1984])

In contrast to its effects in the posterior accumbens, in the anterior accumbens CI-988 did block the inhibition by CCK-8S of DA-induced hyperlocomotion. [Crawley, 1992] In this study, the effects of administration of DA, CCK-8S and CCK receptor antagonists into the VTA on locomotor activity were also determined. After injection into the VTA, CI-988 and L-365,260 blocked the potentiation by CCK-8S of DA-induced hypolocomotion. It was concluded from these data that VTA DA/CCK-8S neurons that project to the posterior accumbens possess CCK-B receptors on their soma and dendrites and CCK-A receptors on their terminals. The former conclusion is consistent with the result of an electrophysiological study in which CCK-8S-potentiation of apomorphine-induced inhibition of A10 DA neuronal firing rate was attributed to CCK-B receptors (see Section IV B.1. above). [Hommer et al., 1986; but see Kelland et al., 1991] This finding may also be related to the fact that subpopulations of A10 DA neurons and identified mesoaccumbal DA neurons are inhibited by CCK-8S, an effect that may be due to CCK-B receptor stimulation (see Section IV A.1. above). [Kelland et al., 1991; Skirboll et al., 1981]

The modulatory effects of intra-accumbal CCK-8S on DA-mediated locomotor activity suggested that CCK-8S may itself play a role in the normal control of locomotion. Intraaccumbal administration of CCK antagonists alone, however, had no effects on either DA-induced changes in locomotion or on dark-induced hyperlocomotion. [Crawley, 1992] Subcutaneous injection of devazepide and L-365,260 rats also did not affect spontaneous locomotor activity in rats. [Higgins et al., 1994] In this study, rats were separated on the basis of differences in their responsiveness to the locomotor-activating effects of amphetamine. L-365,260 potentiated amphetamine-induced hyperlocomotion in rats characterized as "low-amphetamine" responders, whereas the behavior of "high-responders" was unaffected. It was suggested that rats with a low response to amphetamine may have a higher basal CCK-B tone in the nucleus accumbens that diminishes DA activity. In light of the results of the study by Crawley referred to above, this effect occurs presumably in the anterior accumbens.

VII. CLINICAL RELEVANCE OF CCK

A. ROLE OF CCK IN NEUROLOGICAL DISEASE

Nigral CCK levels were found to be decreased in Parkinson's disease [Studler et al., 1982], although a later study did not confirm this finding. [Fernandez et al., 1992] CCK receptors in the caudate and putamen are reduced in Huntington's chorea. [Hays et al., 1981] Nigral levels of CCK are also reduced in this disease. [Emson et al., 1980] Administration of the CCK analogue, ceruletide, to patients with tardive dyskinesia and other neurological disorders reduced choreic movements. [Hashimoto and Yanagisawa, 1990; Matsunaga et al., 1988; Nishikawa et al., 1985]

In monkeys rendered parkinsonian by treatment with the DA neurotoxin MPTP, L-365,260 but not devazepide potentiated DA agonist-induced hyperlocomotion. [Boyce et al., 1990] This result is consistent with the potentiation by L-365,260 of amphetamine-induced locomotion in rats with low responsiveness to amphetamine. [Higgins et al., 1994]

B. ROLE OF CCK IN PSYCHOSIS

CCK-8S exerts neuroleptic-like behavioral effects. [Zetler, 1985] In light of the postulated role of DA systems in psychosis, several investigators have hypothesized that CCK-8 has a role in schizophrenia. [Bunney et al., 1985; Hökfelt et al., 1980; Wang et al., 1984] Clinical studies of CCK-8 and ceruletide (a decapeptide analog of CCK) in schizophrenia have generally yielded negative results. Although several early studies suggested beneficial effects of CCK-8 and ceruletide in schizophrenia, many of these studies did not use a double-blind design or test the drugs in neuroleptic-free patients. [for reviews, see Montgomery and

Green, 1988; Nair et al., 1985] Most controlled studies reported no improvement in neuroleptic-free schizophrenics or schizophrenics maintained on neuroleptics during CCK-8 or ceruletide treatment. [Albus et al., 1986; Hommer et al., 1984; Lotstra et al., 1984; Mattes et al., 1985; Mizuki et al., 1988; Peselow et al., 1987; Tamminga et al., 1986; for review, see Montgomery and Green, 1988] The CCK antagonist proglumide was also without effect in patients receiving neuroleptic treatment. [Innis et al., 1985; Whiteford et al., 1992]

Several investigators have measured the concentration of CCK in the cerebrospinal fluid of schizophrenics. [Gerner et al., 1985; Rafaelsen and Gjerris, 1985; Tamminga et al., 1986; Verbanck et al., 1984] In comparison to controls, these studies found either net increases, decreases or no difference in the level of CCK in the cerebrospinal fluid of schizophrenics. The lack of consensus among these studies may stem from different assay methods, characteristics of the patient populations or other methodological differences. [Beinfeld and Garver, 1991]

In man, low levels of CCK mRNA have been observed in the nucleus paranigralis (human equivalent of the rodent A10 area) and substantia nigra. [Palacios et al., 1989; Schalling et al., 1990] Extensive distribution of CCK mRNA was observed, however, within the midbrain DA neurons of several neuroleptic-treated schizophrenia patients. [Schalling et al., 1990] It is not yet known if these CCK mRNA increases are due to neuroleptic treatment or to the schizophrenic condition.

C. ROLE OF CCK-8S IN ANXIETY AND PANIC

An exciting recent development in CCK research concerns the identification of a role for endogenous CCK-8S in anxiety and panic disorder. [for reviews, see Harro et al., 1993; Ravard and Dourish, 1990] CCK-4 induces panic attacks in patients with panic disorder and in normal controls. [Bradwejn et al., 1990, 1991, 1992, 1994ab; de Montigny, 1989] Panic disorder patients are more sensitive than controls to this effect of CCK-4. [Bradwejn et al., 1991] Pentagastrin was anxiogenic in patients and controls, whereas only patients with panic disorder experienced panic attacks in one study. [Abelson and Nesse, 1994] In another study, patients were far more likely than controls to experience panic in response to pentagastrin. [van Megen et al., 1994] Because CCK-4 and pentagastrin are CCK-B receptor agonists, these findings suggested that panic disorder may be responsive to CCK-B antagonist treatment. Indeed, recent results show that L-365,260 antagonizes the panicogenic effect of CCK-4 in panic disorder patients. [Bradwejn et al., 1994ab] The relationship between anxiety and panic is not clear. The results of one recent clinical test showed that patients with a preexisting fear of anxiety symptoms are no more susceptible than other subjects to experience panic in response to CCK-4 administration. [Koszycki et al., 1993]

CCK peptides have been shown to be anxiogenic in animals and, as with the recent human panic data, the evidence implicates CCK-B receptors in this response. Peripheral and central administration of CCK-8S and CCK-8U was reported to increase fear motivation in rats as assessed by active/passive avoidance paradigms, a result that implicates CCK-B receptors. [Fekete et al., 1984] In the elevated plus-maze paradigm, i.p. administration of the CCK-8S agonist ceruletide produced an angiogenic-like response in mice (decreased exploratory behavior) that was prevented by coadministration of proglumide [Harro et al., 1990] Injection of CCK-8S into the posterior nucleus accumbens of rats also decreased exploratory behavior in this test. [Derrien et al., 1993] Systemic (i.p.) administration of CCK-8U is anxiogenic in the rat elevated plus-maze test, an effect that was attributed to CCK-B receptor stimulation. [Chopin and Briley, 1993] In accordance with the implication of a role for CCK-B receptors in anxiety, s.c. administration of CCK-4 had angiogenic-like effects in the rat elevated plus-maze test, and L-365,260 was far more potent than devazepide and lorglumide in reversing these effects. [Harro and Vasar, 1991] The anxiogenic-like effects of the CCK-B agonist BC197 in the rat elevated plus-maze test were similarly blocked by CI-988. [Derrien et al., 1994]

In some studies, anxiolytic effects have been reported after administration of CCK receptor antagonists. CI-988 and L-365,260 were found to be anxiolytic in the mouse and rat elevated plus-maze tests. [Hughes et al., 1990; Rataud et al., 1991; Singh et al., 1991a] In addition, CI-988 and L-365,260, but not devazepide, were anxiolytic in the mouse light/dark test and the rat social interaction test. [Singh et al., 1991b] CI-988 blocked the anxiogenic effect (as measured by activity in the light/dark box) of withdrawal from diazepam administration in mice [Singh et al., 1992] There is also evidence for anxiolytic effects of CI-988 in monkeys. [Powell and Barrett, 1991; Woodruff and Hughes, 1991] In contrast to studies that implicate a role for CCK-B receptors in anxiety, devazepide was reported to block the effects of CCK-8S in the elevated plus-maze paradigm following their injection into the posterior nucleus accumbens. [Daugé et al., 1989] In addition, another study of the effects of CCK antagonists in the mouse light/dark test found devazepide but not L-365,260 to be anxiolytic. [Hendrie et al., 1993] The authors of the latter report suggest that subtle differences in animal tests of anxiety (e.g., presence of aversive properties of the apparatus, and the particular measurements made) can result in marked differences in the sensitivity of the model.

Brain CCK-4 and CCK-8 levels were measured *postmortem* in a rat model of anxiety where rats are exposed to the smell of a predator (cat). [Pavlasevic et al., 1993] CCK-4 concentrations were increased in several regions including the olfactory bulb, central nucleus of the amygdala, frontal cortex, and the dorsal and ventral striatum. In contrast, CCK-8 levels were increased only in the ventral striatum. These results support a role for CCK-4 and CCK-B receptors in anxiety. Clearly, many of the brain regions that displayed changes in CCK peptide levels in this model of anxiety receive prominent DA inputs (except for the olfactory bulb, which contains intrinsic DA neurons). In this emerging field of study, the influence of DA systems in models of CCK-induced anxiety or panic is sure to receive more attention.

VIII. SIGNIFICANCE OF DA/CCK-8S COLOCALIZATION

Verhage et al. identified several differences between the release characteristics of hippocampal CCK-8S and those of classical transmitters such as glutamate, GABA and catecholamines. [Verhage et al., 1991] These differences concern vesicle type, calcium-sensitivity of the release mechanism, and the site of release. It is not known whether these differences apply to DA/CCK-8S *colocalization*, but it is likely that the physiological state of the DA/CCK-8S neuron determines the magnitude of release of each of these cotransmitters as has been shown for the release of DA and neurotensin from DA/neurotensin nerve terminals in the prefrontal cortex. [Bean and Roth, 1991] In other known instances of classical neurotransmitter/peptide coexistence, the electrophysiological control of peptide release differs from that for the amine, with higher firing frequencies and burst-firing required for peptide release. [Bartfai et al., 1988] DA is released in the striatum and olfactory tubercle during stimulation of the medial forebrain bundle at physiological frequencies; a bursting pattern enhances this release. [Gonon, 1988; Suaud-Chagny et al., 1991] Perhaps bursting is required for *corelease* of DA and CCK-8S. Because DA/CCK-8S colocalization is much greater in the posterior accumbens than in the caudate, comparison of experimental results obtained in these regions will be useful to assess the influence of DA/CCK-8S cotransmission on neuronal activity in DA-innervated regions of the forebrain.

The studies described above demonstrate that, in addition to its existence within DA neurons, CCK-8 has complex electrophysiological effects and biochemical effects on DA neurons. It is now known that DA and CCK-8 modulate each other's somatodendritic release as well as release in DA/CCK-8S terminal regions, and have opposing effects on the firing rate of neurons in the caudate nucleus and nucleus accumbens. This information is essential to support the concept that peptide/monoamine coexistence has functional significance with regard to cotransmission. Much more work is needed to clarify the mechanisms by which CCK-8S and DA interact to influence presynaptic and postsynaptic cell function. For example, it will be important to determine under what conditions endogenous and exogenous CCK-8S influence DA neurotransmission. Studies of this nature will have relevance for current hypotheses about the role of midbrain DA/CCK-8 systems in schizophrenia, Parkinson's disease and other psychiatric and neurological conditions linked to DA neuronal dysfunction.

REFERENCES

Abelson, J.L. and Nesse, R.M., Pentagastrin infusions in patients with panic disorder I. Symptoms and cardiovascular responses, *Biol. Psychiatry*, 36, 73-83, 1994.

Albus, M., von Gellhorn, K., Münch, U., Naber, D. and Ackenheil, M., A double-blind study with ceruletide in chronic schizophrenic patients: biochemical and clinical results, *Psychiatry Res.,* 19, 1-7, 1986.

Altar, C.A. and Boyar, W.C., Brain CCK-B receptors mediate the suppression of dopamine release by cholecystokinin, *Brain Res.,* 483, 321-326, 1989.

Altar, C.A., Boyar, W.C., Oei, E. and Wood, P.L., Cholecystokinin mediates basal and drug-induced increases of limbic and striatal dopamine release, *Brain Res.,* 460, 76-82, 1988.

Anderson, D. R., Li, W. and Tepper, J M., GABAergic inhibition of nigrostriatal dopaminergic neurons by selective activation of pars reticulata projection neurons, *Soc. Neurosci. Abst.,* 19, 1432, 1993.

Artaud, F., Baruch, P., Stutzmann, J.M., Saffroy, M., Dodeheu, G., Barbeito, L., Hervé, D., Studler, J.M., Glowinski, J. and Chéramy, A., Cholecystokinin: corelease with dopamine from nigrostriatal neurons in the cat, *Eur. J. Neurosci.,*162-171, 1989.

Bartfai, T., Iverfeldt, K. and Fisone, G., Regulation of the release of coexisting neurotransmitters, *Ann. Rev. Pharmacol. Toxicol.,* 28, 285-310, 1988.

Bean, A.J. and Roth, R.H., Extracellular dopamine and neurotensin in rat prefrontal cortex *in vivo*: effects of median forebrain bundle stimulation frequency, stimulation pattern, and dopamine autoreceptors, *J. Neurosci,* 11, 2694-2702, 1991.

Beinfeld, M.C., An HPLC and RIA analysis of the cholecystokinin peptides in rat brain, *Neuropeptides,* 1, 203-209, 1981.

Beinfeld, M.C., Cholecystokinin in the central nervous system: a minireview, *Neuropeptides,* 3, 411-427, 1983.

Beinfeld, M.C. and Garver, D.L., Concentration of cholecystokinin in cerebrospinal fluid is decreased in psychosis: relationship to symptoms and drug response, *Prog. Neuro-Psychopharmacol. Biol. Psychiatry,* 15, 601-609, 1991.

Boyce, S., Rupniak, N.M.J., Tye, S., Steventon, M.J. and Iversen, S.D., Modulatory role of CCK-B antagonists in Parkinson's disease, *Clin. Neuropharmacol.,* 13, 339-347, 1990.

Bradwejn, J., Koszycki, D., Annable, L., Couëtoux-du Tertre, A., Reines, S. and Karkanias, C., A dose-ranging study of the behavioral and cardiovascular effects of CCK-tetrapeptide in panic disorder, *Biol. Psychiatry,* 32, 903-912, 1992.

Bradwejn, J., Koszycki, D., Dutertre, A.C., Paradis, M. and Bourin, M., Effects of flumazenil on cholecystokinin tetrapeptide-induced panic symptoms in healthy volunteers, *Psychopharmacology,* 114, 257-261, 1994a.

Bradwejn, J., Koszycki, D., Dutertre, A.C., VanMegen, H., Denboer, J., Westenberg, H. and Annable, L., The panicogenic effects of cholecystokinin-tetrapeptide are antagonized by L-365,260, a central cholecystokinin receptor antagonist, in patients with panic disorder, *Arch. Gen. Psychiatry,* 51, 486-493, 1994b.

Bradwejn, J., Koszycki, D. and Meterissian, G., Cholecystokinin-tetrapeptide induced panic attacks in patients with panic disorder, *Can. J. Psychiatry*, 35, 83-85, 1990.

Bradwejn, J., Koszycki, D. and Shriqui, C., Enhanced sensitivity to cholecystokinin tetrapeptide in panic disorder, *Arch. Gen. Psychiatry*, 48, 603-610, 1991.

Brodie, M.S. and Dunwiddie, T.V., Cholecystokinin potentiates dopamine inhibition of mesencephalic dopamine neurons in vitro, *Brain Res.*, 425, 106-113, 1987.

Brog, J.S. and Beinfeld, M.C., Cholecystokinin release from the rat caudate-putamen, cortex and hippocampus is increased by activation of the D_1 dopamine receptor, *J. Pharmacol. Exp. Ther.*, 260, 343-348, 1992.

Bunney, B.S., The electrophysiological pharmacology of midbrain dopaminergic systems, in *The Neurobiology of Dopamine*, eds., Horn, A.S., Korf, J. and Westerink, B.H.C., Eds., Academic Press, New York, 1979, chap. 25.

Bunney, B.S., Chiodo, L.A. and Freeman, A.S., Further studies on the specificity of proglumide as a selective cholecystokinin antagonist in the central nervous system, *Ann. N.Y. Acad. Sci.*, 448, 345-351, 1985.

Butcher, S.P., Varro, A., Kelly, J.S. and Dockray, G.J., In vivo studies on the enhancement of cholecystokinin release in the rat striatum by dopamine depletion, *Brain Res.*, 505, 119-122, 1989.

Carlberg, M., Gundlach, A.L., Mercer, L.D. and Beart, P.M., Autoradiographic localization of cholecystokinin A and B receptors in rat brain using $[^{123}I]D$-$tyr^{25}(nle^{28, 31})$-CCK 25-33S, *Eur J. Neurosci.*, 4, 563-573, 1992.

Cheramy, A., Leviel, V. and Glowinski, J., Dendritic release of dopamine in the substantia nigra, *Nature*, 289, 537-542, 1981.

Chiodo, L.A., Dopamine-containing neurons in the mammalian central nervous system: electrophysiology and pharmacology, *Neurosci. Biobehav. Rev.*, 12, 49-91, 1988.

Chiodo, L.A., Bannon, M.J., Grace, A.A., Roth, R.H. and Bunney, B.S., Evidence for the absence of impulse-regulating somatodendritic and synthesis-modulating nerve terminal autoreceptors on subpopulations of mesocortical dopamine neurons, *Neuroscience*, 12, 1-16, 1984.

Chiodo, L.A. and Bunney, B.S., Proglumide: selective antagonism of excitatory effects of cholecystokinin in central nervous system, *Science*, 219, 1449-1451, 1983.

Chiodo, L.A. and Bunney, B.S., Population response of midbrain dopaminergic neurons to neuroleptics: further studies on time course and non-dopaminergic neuronal influences, *J. Neurosci.*, 7, 629-633, 1987.

Chiodo, L.A., Freeman, A.S. and Bunney, B.S., Electrophysiological studies on the specificity of the cholecystokinin antagonist proglumide, *Brain Res.*, 410, 205-211, 1987.

Chopin, P. and Briley, M., The benzodiazepine antagonist flumazenil blocks the effects of CCK receptor agonists and antagonists in the elevated plus-maze, *Psychopharmacology,* 110, 409-414, 1993.

Conzelmann, U., Holland, A. and Meyer, D.K., Effects of selective dopamine D_2-receptor agonists on the release of cholecystokinin-like immunoreactivity from rat neostriatum, *Eur. J. Pharmacol.,* 101, 119-125, 1984.

Crawley, J.N., Subtype-selective cholecystokinin receptor antagonists block cholecystokinin modulation of dopamine-mediated behaviors in the rat mesolimbic pathway, *J. Neurosci.,* 12, 3380-3391, 1992.

Crawley, J.N., Hommer, D.W. and Skirboll, L.R., Topographical analysis of nucleus accumbens sites at which cholecystokinin potentiates dopamine-induced hyperlocomotion in the rat, *Brain Res.,* 35, 337-341, 1985a.

Crawley, J.N., Kiss, J.Z. and Mezey, E., Bilateral midbrain transections block the behavioral effects of cholecystokinin on feeding and exploration in rats, *Brain Res.,* 322, 316-321, 1984.

Crawley, J.N., Stivers, J.A., Blumstein, L.K. and Paul, S.M., Cholecystokinin potentiates dopamine-mediated behaviors: evidence for modulation specific to a site of coexistence, *J. Neurosci.,* 5, 1972-1983, 1985b.

Crawley, J.N., Stivers, J.A., Hommer, D.W., Skirboll, L.R. and Paul, S.M., Antagonists of central and peripheral behavioral actions of cholecystokinin octapeptide, *J. Pharmacol. Exp. Ther.,* 236, 320-330, 1986.

Debonnel, G. and de Montigny, C., Increased neuronal responsiveness to cholecystokinin and dopamine induced by lesioning mesolimbic dopaminergic neurons: an electrophysiological study in the rat, *Synapse,* 2, 537-545, 1988.

de Montigny, C., Cholecystokinin tetrapeptide induces panic-like attacks in healthy volunteers, *Arch. Gen. Psychiatry,* 46, 511-517, 1989.

Deniau, J.M., Kitai, S.T., Donoghue, J.P. and Grofova, I., Neuronal interactions in the substantia nigra pars reticulata through axon collaterals of the projection neurons, *Exp. Brain Res.,* 47, 105-113, 1982.

Derrien, M., McCort-Tranchepain, Ducos, B., Roques, B.P. and Durieux, C., Heterogeneity of CCK-B receptors involved in animal models of anxiety, *Pharmacol. Biochem. Behav.,* 49, 133-141, 1994.

De Witte, Ph., Heidbreder, C., Roques, B. and Vanderhaeghen, J.-J., Opposite effects of cholecystokinin octapeptide (CCK-8) and tetrapeptide (CCK-4) after injection into the caudal part of the nucleus accumbens or into its rostral part and the cerebral ventricles, *Neurochem. Intl.,* 10, 473-479, 1987.

Dockray, G.J., Cholecystokinin in rat cerebral cortex, *Brain Res.,* 188, 155-165, 1980.

Emson, P.C., Lee, C.M. and Rehfeld, J.F., Cholecystokinin octapeptide: vesicular localization and calcium dependent release from rat brain in vitro, *Life Sci.,* 26, 2157-2163, 1980.

Emson, P.C., Rehfeld, J.F., Langevin, H. and Rossor, M., Reduction in cholecystokinin-like immunoreactivity in the basal ganglia in Huntington's disease, *Brain Res.*, 198, 497-500, 1980.

Fallon, J.H., Hicks, R. and Loughlin, S.E., The origin of cholecystokinin terminals in the basal forebrain of the rat: evidence from immunofluorescence and retrograde tracing, *Neurosci. Lett.*, 37, 29-35, 1983.

Fekete, M., Lengyel, A., Hegedüs, B., Penke, B., Zarándy, M., Tóth, G.K. and Telegdy, G., Further analysis of the effects of cholecystokinin octapeptides on avoidance behaviour in rats, *Eur. J. Pharmacol.*, 98, 79-91, 1984.

Fernandez, A., de Ceballos, M.L., Jenner, P. and Marsden, C.D., Striatal neuropeptide levels in Parkinson's disease, *Neurosci. Lett.*, 145, 171-174, 1992.

Freeman, A.S. and Bunney, B.S., Activity of A9 and A10 dopaminergic neurons in unrestrained rats: further characterization and effects of apomorphine and cholecystokinin, *Brain Res.*, 405, 46-55, 1987.

Freeman, A.S. and Chiodo, L.A., Electrophysiological aspects of Cholecystokinin/dopamine interactions, in *Neurophysiology of Dopaminergic Systems: Current Status and Clinical Perspectives*, eds., Chiodo, L.A. and Freeman, A.S., Eds., Lakeshore Publishing, Grosse Pointe, MI, 1987, chap. VIII.

Freeman, A.S. and Chiodo, L.A., Electrophysiological effects of cholecystokinin octapeptide on identified rat nigrostriatal neurons, *Brain Res.*, 439, 266-274, 1988.

Freeman, A.S., Chiodo, L.A., Lentz, S.I., Wade, K. and Bannon, M.J., Release of CCK from midbrain slices and modulatory influence of D2 DA receptor stimulation, *Brain Res.*, 555, 281-287, 1991.

Freeman, A.S., Henry, D.J. and Chiodo, L.A., Effects of quinpirole and cholecystokinin octapeptide on dopamine neuronal activity in the midbrain slice, *Soc. Neurosci. Abst.*, 15, 1000, 1989.

Freeman, A.S., Shen, R.-Y. and Chiodo, L.A., Interactions of CCK and CCK antagonists with midbrain dopaminergic neurons: electrophysiological aspects, in *CCK Antagonists*, Wang, R.Y. and Schoenfeld, R., Eds., Liss, New York, pp.181-198, 1988.

Freeman, A.S. and Zhang, J., Electrophysiological effects of CCK peptides on substantia nigra pars reticulata neurons, *Soc. Neurosci. Abst.*, 19, 1373, 1993.

Gerhardt, G.A., Friedemann, M., Brodie, M.S., Vickroy, T.W.,Gratton, A.P., Hoffer, B.J. and Rose, G.M., The effects of cholecystokinin on dopamine-containing nerve terminals in the caudate nucleus and nucleus acumbens of the anesthetized rat: an in vivo electrochemical study, *Brain Res.*, 499, 157-163, 1989.

Gerner, R.H., Vankammen, D.P. and Ninan, P.T., Cerebrospinal fluid cholecystokinin, bombesin and somatostatin in schizophrenia and normals, *Prog. Neuro-Psychopharmacol. Biol. Psychiatry*, 9, 73-82, 1985.

Gilles, C., Lotstra, F. and Vanderhaeghen, J.J., CCK nerve terminals in the rat striatal and limbic areas originate partly in the brain stem and partly in telencephalic structures, *Life Sci.*, 32, 1683-1690, 1983.

Gonon, F.G., Nonlinear relationship between impulse flow and dopamine released by rat midbrain dopaminergic neurons as studies by in vivo electrochemistry, *Neuroscience*, 24, 19-28, 1988.

Grace, A.A. and Bunney, B.S., Paradoxical GABA excitation of nigral dopaminergic cells: indirect mediation through reticulata inhibitory neurons, *Eur. J. Pharmacol.*, 59, 211-218, 1979.

Graham, W.C., Hill, D.R., Woodruff, G.N., Sambrook, M.A. and Crossman, A.R., Reduction of [^{125}I]Bolton Hunter CCK8 and [^{3}H]MK-329 (devazepide) binding to CCK receptors in the substantia nigra/VTA complex and its forebrain projection areas following MPTP-induced hemi-parkinsonism in the monkey, *Neurosci. Lett.*, 131, 129-134, 1991.

Hamilton, M.E. and Freeman, A.S., Influence of intra-VTA CCK peptides on DA overflow in nucleus accumbens and amygdala in freely moving rats, *Soc. Neurosci. Abst.*, 20, 286, 1994.

Hamilton, M.E. and Freeman, A.S., Effects of administration of cholecystokinin into the VTA on DA overflow in nucleus accumbens and amygdala of freely moving rats, *Brain Res.*, 688, 134-142, 1995.

Harro, J., Põld, M. and Vasar, E., Anxiogenic-like action of caerlein, a CCK-8 receptor agonist, in the mouse: influence of acute and subchronic diazepam treatment, *Naunyn-Schmiedeberg's Arch. Pharmacol.*, 341, 62-67, 1990.

Harro, J. and Vasar, E., Evidence that CCK$_B$ receptors mediate the regulation of exploratory behavior in the rat, *Eur. J. Pharmacol.*, 193, 379-381, 1991.

Harro, J, Vasar, E. and Bradwejn, J., CCK in animal and human research on anxiety, *Trends Pharmacol. Sci.*, 14, 244-249, 1993.

Hashimoto, T. and Yanagisawa, N., Acute reduction and long-term improvement of chorea with ceruletide (cholecystokinin analogue), *J. Neurol. Sci.*, 100, 178-185, 1990.

Hays, S.E., Goodwin, F.K. and Paul, S.M., Cholecystokinin receptors are decreased in basal ganglia and cerebral cortex of Huntington's disease, *Brain Res.*, 225, 452-456, 1981.

Hendrie, C.A., Neill, J.C., Shepherd, J.K. and Dourish, C.T., The effects of CCK$_A$ and CCK$_B$ antagonists on activity in the black/white exploration model of anxiety in mice, *Physiol. Behav.*, 54, 689-693, 1993.

Hill, D.R., Campbell, N.J., Shaw, T.M. and Woodruff, G.N., Autoradiographic localization and biochemical characterization of,peripheral type CCK receptors in rat CNS using highly selective nonpeptide CCK antagonists, *J. Neurosci.*, 7, 2967-2976, 1987.

Hill, D.R., Shaw, T.M., Dourish, C.T. and Woodruff, G.N., CCK-A receptors in the rat interpeduncular nucleus: evidence for a presynaptic location, *Brain Res.*, 454, 101-105, 1988.

Hill, D.R., Shaw, T.M., Graham, W. and Woodruff, G.N., Autoradiographical detection of cholecystokinin-A receptors in primate brain using ^{125}I-Bolton Hunter CCK-8 and ^{3}H-MK-329, *J.Neurosci.*, 10, 1070-1081, 1990.

Hökfelt, T., Skirboll, L., Rehfeld, J.F., Goldstein, M., Markey, K. and Dann, O., A subpopulation of mesencephalic dopamine neurons projecting to limbic areas contains a cholecystokinin-like peptide, *Neuroscience*, 5, 2093-2124, 1980.

Hommer, D.W., Palkovits, M., Crawley, J.N., Paul, S.M. and Skirboll, L.R., Cholecystokinin-induced excitation in the substantia nigra: evidence for peripheral and central components, *J. Neurosci.*, 6, 1387-1392, 1985.

Hommer, D.W., Pickar, D., Roy, A., Ninan, P., Boronow, J. and Paul, S.M., The effects of ceruletide in schizophrenia, *Arch. Gen. Psychiatry*, 41, 617-619, 1984.

Hommer, D.W. and Skirboll, L.R., Cholecystokinin-like peptides potentiate apomorphine-induced inhibition of dopamine neurons, *Eur. J. Pharmacol.*, 91, 151-152, 1983.

Hommer, D.W., Stoner, G., Crawley, J.N., Paul, S.M. and Skirboll, L.R., Cholecystokinin-dopamine coexistence: electrophysiological actions corresponding to cholecystokinin receptor subtype, *J. Neurosci.*, 6, 3039-3043, 1986.

Honda, T., Wada, E., Battey, J.F. and Wank, S.A., Differential gene expression of CCK_A and CCK_B receptors in the rat brain, *Mol. Cell. Neurosci.*, 4, 143-154, 1993.

Hughes, J., Boden, P., Costall, B., Domeney, A., Kelly, E., Horwell, D.C., Hunter, J.C., Pinnock, R.D. and Woodruff, G.N., Development of a class of selective cholecystokinin type B receptor antagonists having potent anxiolytic activity, *Proc. Natl. Acad. Sci. USA*, 87, 6728-6732, 1990.

Hurd, Y.L., Lindefors, N., Brodin, E., Brené, S., Persson, H., Ungerstedt, U. and Hökfelt, T., Amphetamine regulation of mesolimbic dopamine/ cholecystokinin neurotransmission, *Brain Res.*, 578, 317-326, 1992.

Hutchison, J.B., Strupish, J. and Nahorski, S.R., Release of endogenous dopamine and cholecystokinin from rat striatal slices: effects of amphetamine and dopamine antagonists, *Brain Res.*, 370, 310-314, 1986.

Innis, R.B., Bunney, B.S., Charney, D.S., Price, L.H., Glazer, W.M., Sternberg, D.E., Rubin, A.L. and Heninger, G.R., Does the cholecystokinin antagonist proglumide possess antipsychotic activity?, *Psychiatry Res.*, 18, 1-7, 1985.

Innis, R.B. and Snyder, S.H., Distinct cholecystokinin receptors in brain and pancreas, *Proc. Natl. Acad. Sci. USA*, 77, 6917-6921, 1980.

Ito, M., Matsui, T., Taniguchi, T., Tsukamoto, T., Murayama, T., Arima, N., Nakata, H., Chiba, T. and Chihara, K., Functional characterization of a human brain cholecystokinin-B receptor: a trophic effect of cholecystokinin and gastrin, *J. Biol. Chem.*, 268, 18300-18305, 1993.

Jiang, L.H., Kasser, R.J. and Wang, R.Y., Cholecystokinin antagonist lorglumide reverses chronic haloperidol-induced effects on dopamine neurons, *Brain Res.*, 473, 165-168, 1988.

Kalivas, P.W. and Duffy, P., A comparison of axonal and somatodendritic dopamine release using in vivo dialysis, *J. Neurochem.*, 56, 961-967, 1991.

Kariya, K., Tanaka, J. and Nomura, M., Systemic administration of CCK-8S, but not CCK-4, enhances dopamine turnover in the posterior nucleus accumbens: a microdialysis study in freely moving rats, *Brain Res.*, 657, 1-6, 1994.

Kelland, M.D., Zhang, J., Chiodo, L.A. and Freeman, A.S., Receptor selectivity of cholecystokinin effects on mesoaccumbens dopamine neurons, *Synapse*, 8, 137-143, 1991.

Kihara, T., Ikeda, M., Ibii, N. and Matsushita, A., Ceruletide, a CCK-like peptide, attenuates dopamine release from the rat striatum via a central site of action, *Brain Res.*, 588, 270-276, 1992.

Koszycki, D., Cox, B.J. and Bradwejn, J., Anxiety sensitivity and response to cholecystokinin tetrapeptide in healthy volunteers, *Am. J. Psychiatry*, 150, 1881-1883, 1993.

Ladurelle, N., Keller, G., Roques, B.P. and Daugé, V., Effects of CCK_8 and of the CCK_B-selective agonist BC264 on extracellular dopamine content in the anterior and posterior nucleus accumbens: a microdialysis study in freely moving rats, *Brain Res.*, 628, 254-262, 1993.

Laitinen, K., Crawley, J.N., Mefford, I.N. and De Witte, Ph., Neurotensin and cholecystokinin microinjected in to the ventral tegmental area modulate microdialysate concentrations of dopamine and metabolites in the posterior nucleus accumbens, *Brain Res.*, 523, 342-346, 1990.

Lane, R.F., Blaha, C.D. and Phillips, A.G., In vivo electrochemical analysis of cholecystokinin-induced inhibition of dopamine release in the nucleus accumbens, *Brain Res.*, 397, 200-204, 1986.

Lee, Y.-M., Beinborn, M., McBride, E.W., Lu, M., Kolakowski, L.F. Jr. and Kopin, A.S., The human brain cholecystokinin-B/gastrin receptor: cloning and characterization, *J. Biol. Chem.*, 268, 8164-8169, 1993.

Liang, R.Z., Wu, M., Yim, C.C.Y. and Mogenson, G.J., Effects of dopamine agonists on excitatory inputs to nucleus accumbens neurons from the amygdala: modulatory actions of cholecystokinin, *Brain Res.*, 554, 85-94, 1991.

Llinas, R., Greenfield, S.A. and Jahnsen, H., Electrophysiology of pars compacta cells in the in vitro substantia nigra - a possible mechanism for dendritic release, *Brain Res.*, 294, 127-132, 1984.

Lotstra, F., Verbanck, P., Mendlewicz, J. and Vanderhaeghen, J.J., No evidence of antipsychotic effect of caerulin in schizophrenic patients free of neuroleptics: a double blind crossover study, *Biol. Psychiatry*, 19, 877-882, 1984.

Markstein, R. and Hökfelt, T., Effect of cholecystokinin-octapeptide on dopamine release from slices of cat caudate nucleus, *J. Neurosci.*, 4, 570-575, 1984.

Marley, P.D., Emson, P.C. and Rehfeld, J.F., Effect of 6-hydroxydopamine lesions of the medial forebrain bundle on the distribution of cholecystokinin in rat forebrain, *Brain Res.*, 252, 382-385, 1982.

Marshall, F.H., Barnes, S., Hughes, J., Woodruff, G.N. and Hunter, J.C., Cholecystokinin modulates the release of dopamine from the anterior and posterior nucleus accumbens by two different mechanisms, *J. Neurochem.*, 56, 917-922, 1991.

Marshall, F.H., Barnes, S., Pinnock, R.D. and Hughes, J., Characterization of cholecystokinin octapeptide-stimulated endogenous dopamine release from rat nucleus accumbens in vitro, *Br. J. Pharmacol.*, 99, 845-848, 1990.

Martin, J.R., Beinfeld, M.C. and Wang, R.Y., Modulation of cholecystokinin release from posterior nucleus accumbens by D_2 dopamine receptor, *Brain Res.*, 397, 253-258, 1986.

Massey, B.W., Vanover, K.E. and Woolverton, W.L., Effects of cholecystokinin antagonists on the discriminative stimulus effects of cocaine in rats and monkeys, *Drug Alcohol Res.*, 34:105-111, 1994.

Matsunaga, T., Ohyama, S., Tekehara, S., Kabashima, K., Moriyama, S., Tsuzuki, J., Ikeda, H., Suematsu, M., Akizuki, K. and Fujimoto, K., The effects of ceruletide on tardive dyskinesia: a double-blind placebo-controlled study, *Prog. Neuro-Psychopharmacol. Biol. Psychiatry*, 12, 533-539, 1988.

Mattes, J.A., Hom, W., Rochford, J.M. and Orlosky, M., Ceruletide for schizophrenia: a double-blind study, *Biol. Psychiatry*, 20, 533-538, 1985.

Meltzer, L.T., Christoffersen, C.L., Serpa, K.A. and Razmpour, A., Comparison of the effects of the cholecystokinin-B receptor antagonist, PD 1343408, and the cholecystokinin-A receptor antagonist, L-364,718, on dopamine neuronal activity on the substantia nigra and ventral tegmental area, *Synapse*, 3, 117-122, 1993.

Meyer, D.K., Beinfeld, M.C., Oertel, W.H. and Brownstein, M.J., Origin of the cholecystokinin-containing fibers in the rat caudatoputamen, *Science*, 215, 187-188, 1982.

Meyer, D.K., Holland, A. and Conzelmann, U., Dopamine D_1-receptor stimulation reduces neostriatal cholecystokinin release, *Eur. J. Pharmacol.*, 104, 387-388, 1984.

Meyer, D.K. and Krauss, J., Dopamine modulates cholecystokinin release in neostriatum, *Nature*, 301, 338-340, 1983.

Meyer, D.K., Schultheiss, K. and Hardung, M., Bilateral ablation of frontal cortex reduces concentration of cholecystokinin-like immunoreactivity in rat dorsolateral striatum, *Brain Res.*, 452, 113-117, 1988.

Minabe, Y., Ashby, C.R., Jr. and Wang, R.Y., The CCK-A receptor antagonist devazepide but not the CCK-B receptor antagonist L-365,260 reverses the effects of chronic clozapine and haloperidol on midbrain dopamine neurons, *Brain Res.*, 549, 151-154, 1991.

Mizuki, Y., Ushijima, I., Habu, K., Nakamura, K. and Yamada, M., Effects of ceruletide on clinical symptoms and EEGs in schizophrenia, *Prog. Neuropsychopharmacol. Biol. Psychiatry*, 12, 511-522, 1988.

Montgomery, S.A. and Green, M.C.D., The use of cholecystokinin in schizophrenia: a review, *Psychol. Med.*, 18, 593-603, 1988.

Moran, T.H., Robinson, P.H., Goldrich, M.S. and McHugh, P.R., Two brain cholecystokinin receptors: implications for behavioral actions, *Brain Res.*, 362, 175-179, 1986.

Moran, T.H., Norgren, R., Crosby, R.J. and McHugh, P.R., Central and peripheral vagal transport of cholecystokinin binding sites occurs in afferent fibers, *Brain Res.*, 526, 95-102, 1990.

Morino, P., Mascagni, F., McDonald, A. and Hökfelt, T., Cholecystokinin corticostriatal pathway in the rat - evidence for bilateral origin from medial prefrontal cortical areas, *Neuroscience*, 59, 939-952, 1994.

Nair, N.P.V., Lal, S. and Bloom, D.M., Cholecystokinin peptides, dopamine and schizophrenia - a review, *Prog. Neuropsychopharmacol. Biol. Psychiatry*, 9, 515-524, 1985.

Nishikawa, T., Tanaka, M., Koga, I. and Uchida, Y., Biphasic and long-lasting effect of ceruletide on tardive dyskinesia, *Psychopharmacology*, 86, 43-44, 1985.

Nissbrandt, H., Sundstrom, E., Jonsson, G., Hjorth, S. and Carlsson, A., Synthesis and release of dopamine in rat brain: comparison between substantia nigra pars compacta, pars reticulata and striatum, *J. Neurochem.*, 52, 1170-1182. 1989.

O'Neill, M.F., Dourish, C.T. and Iversen, S.D., Hypolocomotion induced by peripheral or central injection of CCK in the mouse is blocked by the CCK_A receptor antagonist devazepide but not by the CCK_B receptor antagonist L-365,260, *Eur. J. Pharmacol.*, 193, 203-208, 1991.

Palacios, J.M., Savasta, M. and Mengod, G., Does cholecystokinin colocalize with dopamine in the human substantia nigra, *Brain Res.*, 488, 369-375, 1989.

Pavlasevic, S., Bednar, I., Qureshi, G.A. and Södersten, P., Brain cholecystokinin tetrapeptide levels are increased in a rat model of anxiety, *Neuroreport*, 5, 225-228, 1993.

Peselow, E., Angrist, B., Sudilovsky, A., Corwin, J., Siekierski, J., Trent, F. and Rotrosen, J., Double blind controlled trials of cholecystokinin octapeptide in neuroleptic-refractory schizophrenia, *Psychopharmacology*, 91, 80-84, 1987.

Pettit, H.O. and Mueller, K., Infusions of cholecystokinin octapeptide into the ventral tegmental area potentiate amphetamine conditioned place preferences, *Psychopharmacology*, 99, 423-426, 1989.

Powell, K.R. and Barrett, J.E., Evaluation of the effects of PD134308 (CI-988), a CCK-B antagonist, on the punished responding of squirrel monkeys, *Neuropeptides*, 9, 75-78, 1991.

Rafaelsen, O.J. and Gjerris, A., Neuropeptides in the cerebrospinal fluid (CSF) in psychiatric disorders, *Prog. Neuro-Psychopharmacol. Biol. Psychiatry*, 9, 533-538, 1985.

Rasmussen, K., Czachura, J.F., Stockton, M.E. and Howbert, J.J., Electrophysiological effects of diphenylpyrazolidinone cholecystokinin-B and cholecystokinin-A antagonists on midbrain dopamine neurons, *J. Pharmacol. Exp. Ther.*, 64, 480-488, 1993a.

Rasmussen, K., Howbert, J.J. and Stockton, M.E., Inhibition of A9 and A10 dopamine cells by the cholecystokinin-B antagonist LY262691: mediation through feedback pathways from forebrain sites, *Synapse*, 15, 95-103, 1993b.

Rasmussen, K., Stockton, M.E., Czachura, J.F. and Howbert, J.J., Cholecystokinin and schizophrenia: the selective CCK-B antagonist LY 262691 decreases midbrain dopamine unit activity, *Eur. J. Pharmacol.*, 209, 135-138, 1991.

Rasmussen, K., Yu, M.J. and Czachura, J.F., Quinazolinone cholecystokinin (CCK)-B antagonists decrease midbrain dopamine unit activity, *Synapse*, 17, 278-282, 1994.

Rataud, J., Darche, F., Piot, O., Stutzmann, J.-M., Böhme, G.A. and Blanchard, J.-C., 'Anxiolytic' effect of CCK-antagonists on plus-maze behavior in mice, *Brain Res.*, 548, 315-317, 1991.

Ravard, S. and Dourish, C.T., Cholecystokinin and anxiety, *Trends Pharmacol. Sci.*, 11, 271-273, 1990.

Rehfeld, J.F., Immunochemical studies on cholecystokinin. II. Distribution and molecular heterogeneity in the central system and small intestine of man and hog, *J. Biol. Chem.*, 253, 4022-4030, 1978.

Rompré, P.-R. and Boye, S.M., Opposite effects of mesencephalic microinjections of cholecystokinin octapeptide and neurotensin-(1-13) on brain stimulation reward, *Eur. J. Pharmacol.*, 232, 299-303, 1993.

Rouillard, C., Chiodo, L.A. and Freeman, A.S., Electrophysiological effects of cholecystokinin on amygdala neurons, *Soc. Neurosci. Abst.*, 15, 1000, 1989.

Ruggeri, M., Ungerstedt, U., Agnati, L.F., Mutt, V., Härfstrand and Fuxe, K., Effects of cholecystokinin peptides and neurotensin on dopamine release and metabolism in the rostral and caudal part of the nucleus accumbens using intracerebral dialysis in the anaesthetized rat, *Neurochem. Intl.*, 10, 509-520, 1987.

Santiago, M. and Westerink, B.H.C., Characterization and pharmacological responsiveness of dopamine release recorded by microdialysis in the substantia nigra of conscious rats, *J. Neurochem.*, 57, 738-747, 1991.

Savasta, M., Ruberte, E., Palacios, J.M. and Mengod, G., The colocalization of cholecystokinin and tyrosine hydroxylase mRNAs in mesencephalic dopaminergic neurons in the rat brain examined by in situ hybridization, *Neuroscience*, 29, 363-369, 1989.

Schade, R., Vick, K., Sohr, R., Ott, T., Pfister, C., Bellach, J., Mattes, A. and Lemmer, B., Correlative circadian rhythms of cholecystokinin and dopamine content in nucleus accumbens and striatum of rat brain, *Behav. Brain Res.*, 59, 211-214, 1993.

Schalling, M., Friberg, K., Seroogy, K., Riederer, P., Bird, E., Schifferman, S.N., Mailleux, P., Vanderhaeghen, J.J., Kuga, S.,Goldstein, M., Kitahama, K., Luppi, P.H., Jouvet, M. and Hökfelt,T., Analysis of expression of cholecystokinin in dopamine cells in the ventral mesencephalon of several species and in humans with schizophrenia, *Proc. Natl. Acad. Sci. USA*, 87, 8427-8431, 1990.

Seroogy, K.B., Dangaran, K., Lim, S., Haycock, J.W. and Fallon, J.H., Ventral mesencephalic neurons containing both cholecystokinin- and tyrosine hydroxylase-like immunoreactivities project to forebrain regions, *J. Comp. Neurol.*, 279, 397-414, 1989a.

Seroogy, K., Schalling, M., Brene, S., Dagerlind, A., Chai, S.Y., Hökfelt, T., Persson, H., Brownstein, M., Huan, R., Dixon, J., Filer, D., Schlessinger, D. and Goldstein, M., Cholecystokinin and tyrosine hydroxylase messenger RNAs in neurons or rat mesencephalon: peptide/monoamine coexistence studies using in situ hybridization combined with immunocytochemistry, *Exp. Brain Res.*, 74, 149-162, 1989b.

Singh, L., Field, M.J., Hughes, J., Menzies, R., Oles, R.J., Vass, C.A. and Woodruff, G.N., The behavioral properties of CI-988, a selective cholecystokinin$_B$ receptor antagonist, *Br. J. Pharmacol.*, 104, 239-245, 1991a.

Singh, L., Field, M.J., Vass, C.A., Hughes, J. and Woodruff, G.N., The antagonism of benzodiazepine withdrawal effects by the selective cholecystokinin-B receptor antagonist CI-988, *Br. J. Pharmacol.*, 105, 8-10, 1992.

Singh, L., Lewis, A.S., Field, M.J., Hughes, J. and Woodruff, G.N., Evidence for an involvement of the brain cholecystokinin B receptor in anxiety, *Proc. Natl. Acad. Sci. USA*, 88, 1130-1133, 1991b.

Skirboll, L., Grace, A.A., Hommer, D.W., Rehfeld, J., Goldstein, M., Hökfelt, T. and Bunney, B.S., Peptide-monoamine coexistence: Studies of the actions of cholecystokinin-like peptide on the electrical activity of midbrain dopamine neurons, *Neuroscience*, 6, 2111-2124, 1981.

Snyder, G.L., Fisone, G., Morino, P., Gundersen, V., Ottersen, O.P., Hökfelt, T. and Greengard, P., Regulation by the neuropeptide cholecystokinin (CCK-8S) of protein phosphorylation in the neostriatum, *Proc. Natl. Acad. Sci. USA*, 90, 11277-11281, 1993.

Spealman, R.D., Failure of cholecystokinin antagonists to modify the discriminative stimulus effects of cocaine, *Pharmacol. Biochem. Behav.*, 43, 613-616, 1992.

Starr, M.S., Influence of peptides on ³H-dopamine release from superfused rat striatal slices, *Neurochem, Intl.*, 4, 233-240, 1982.

Stittsworth, J.D. and Mueller, A.L., Cholecystokinin octapeptide potentiates the inhibitory response mediated by D2 dopamine receptors in slices of the ventral tegmental area of the brain in the rat, *Neuropharmacology*, 29, 119-127, 1990.

Studler, J.M., Javoy-Agid, F., Cesselin, F., Legrand, J.C. and Agid, Y., CCK-8-immunoreactivity distribution in human brain: selective decrease from the substantia nigra from parkinsonian patients, *Brain Res.*, 243, 176-179, 1982.

Studler, J.M., Simon, H., Cesselin, F., Legrand, J.C., Glowinski, J. and Tassin, J.P., Biochemical investigation on the localization of the cholecystokinin octapeptide in dopaminergic neurons originating from the ventral tegmental area of the rat, *Neuropeptides*, 2, 131-139, 1981.

Suaud-Chagny, M.F., Chergui, K., Chouvet, G. and Gonon, F., Relationship between dopamine release in the rat nucleus accumbens and the discharge activity of dopaminergic neurons during local in vivo application of amino acids in the ventral tegmental area, *Neuroscience*, 49, 63-72, 1992.

Takita, M., Tsurata, T., Oh-hashi, Y. and Kato, T., In vivo release of cholecystokinin-like immunoreactivity in rat frontal cortex under freely moving conditions, *Neurosci. Lett.*, 100, 249-253, 1989.

Tamminga, C.A., Littman, R.L., Alphs, L.D., Chase, T.N., Thaker, G.K. and Wagman, A.M., Neuronal cholecystokinin and schizophrenia: pathologic and therapeutic studies, *Psychopharmacology*, 88, 387-391, 1986.

Tanaka, T., Vincent, S.R., Nomikos, G.G. and Fibiger, H.C., Effect of quinine on autoreceptor-regulated dopamine release in the rat striatum, *J. Neurochem.*, 59, 1640-1645, 1992.

Vaccarino, F.J., Nucleus accumbens dopamine-CCK interactions in psychostimulant reward and related behaviors, *Neurosci. Biobehav. Rev.*, 18, 207-214, 1994.

Vaccarino, F.J. and Vaccarino, J., Nucleus accumbens cholecystokinin (CCK) can either attenuate or potentiate amphetamine-induced locomotor activity: evidence for rostral-caudal differences in accumbens CCK function, *Behav. Neurosci.*, 103, 831-836, 1989.

Vaccarino, F.J. and Vaccarino, A.L., Antagonism of cholecystokinin function in the rostral and caudal nucleus accumbens: differential effects on brain stimulation reward, *Neurosci. Lett.*, 97, 151-156, 1989.

Vanderhaeghen, J.J., Neuronal cholecystokinin, in *Handbook of Chemical Neuroanatomy*, Vol. 4, Part 1, Bjorklund, A. and Hökfelt, T., Eds., Elsevier, New York, 1985, pp. 406-435.

van Megen, H.J.G.M., Westenberg, H.G.M., den Boer, J.A., Haigh, J.R.M. and Traub, M., Pentagastrin induced panic attacks: enhanced sensitivity in panic disorder patients, *Psychopharmacology*, 114, 449-455, 1994.

Van Ree, J.M., Gaffori, O. and DeWied, D., In rats, the behavioral profile of CCK-8 related peptides resembles that of antipsychotic agents, *Eur. J. Pharmacol.*, 93, 63-78, 1983.

Verbanck, P.M.P., Lotstra, F., Gilles, C., Linkowski, P., Mendlewicz, J. and Vanderhaeghen, J.J., Reduced cholecystokinin immunoreactivity in the cerebrospinal fluid of patients with psychiatric disorders, *Life Sci.*, 34, 67-72, 1984.

Verhage, M., Ghijsen, W.E.J.M., Nicholls, D.G. and Wiegant, V.M., Characterization of the release of cholecystokinin-8 from isolated nerve terminals and comparison with exocytosis of classical transmitters, *J. Neurochem.*, 56, 1394-1400, 1991.

Vickroy, T.W., Bianchi, B.R., Kerwin, J.F., Jr., Kopecka, H. and Nadzan, A.M., Evidence that type A CCK receptors facilitate dopamine efflux in rat brain, *Eur. J. Pharmacol.*, 152, 371-372,1988.

Voigt, M.M. and Wang, R.Y., In vivo release of dopamine in the nucleus accumbens of the rat: modulation by cholecystokinin, *Brain Res.*, 296, 189-193, 1984.

Voigt, M.M., Wang, R.Y. and Westfall, T.C., The effects of cholecystokinin on the in vivo release of newly synthesized [^{3}H]dopamine from the nucleus accumbens of the rat, *J. Neurosci.*, 5, 2744-2749, 1985.

Voigt, M.M., Wang, R.Y. and Westfall, T.C., Cholecystokinin octapeptides alter the release of endogenous dopamine from the rat nucleus accumbens in vitro, *J. Pharmacol. Exp. Ther.*, 237, 147-153, 1986.

Wang, R.Y. and Hu, X.-T., Does cholecystokinin potentiate dopamine action in the nucleus accumbens, *Brain Res.*, 380, 363-367, 1986.

Wang, R.Y., Kasser, R.J. and Hu, X.-T., Cholecystokinin receptor subtypes in the rat nucleus accumbens, in *Cholecystokinin Antagonists*, Wang, R.Y. and Schoenfeld, R., Eds., Liss, New York, 1988, pp.199-215.

Wang, R.Y., White, F.J. and Voigt, M.M., Cholecystokinin, dopamine and schizophrenia, *Trends Pharmacol. Sci.*, 5, 436-438, 1984.

Wank, S.A., Harkins, R., Jensen, R.T., Shapira, H., de Weerth, A. and Slattery, T., Purification, molecular cloning, and functional expression of the cholecystokinin receptor from rat pancreas, *Proc. Natl. Acad. Sci. USA*, 89, 3125-3129, 1992.

Wank, S.A., Pisegna, J.R. and de Weerth, A., Brain and gastrointestinal cholecystokinin receptor family: structure and functional expression, *Proc. Natl. Acad. Sci. USA*, 89, 8691-8695, 1992.

Waszczak, B.L., Lee, E.K., Tamminga, C.A. and Walters, J.R., Effect of dopamine system activation on substantia nigra pars reticulata output neurons: variable single-unit responses in normal rats and inhibition in 6-hydroxydopamine-lesioned rats, *J. Neurosci.*, 4, 2369-2375, 1984.

Waszczak, B.L. and Walters, J.R., Intravenous GABA agonist administration stimulates firing of A_{10} dopaminergic neurons, *Eur. J. Pharmacol.*, 66, 141-144, 1980.

Waszczak, B. L. and Walters, J. R., Endogenous dopamine can modulate inhibition of substantia nigra pars reticulata neurons elicited by GABA iontophoresis or striatal stimulation, *J. Neurosci.*, 6, 120-126, 1986.

Weiss, F., Tanzer, D.J. and Ettenberg, A., Opposite actions of CCK-8 on amphetamine-induced hyperlocomotion and stereotypy following intracerebroventricular and intra-accumbens injections in rats, *Pharmacol. Biochem. Behav.*, 30, 309-317, 1988.

Wettstein, J.G., Buéno, L. and Junien, J.L., CCK antagonists: pharmacology and therapeutic interest, *Pharmacol. Ther.*, 62, 267-282, 1994.

White, F.J. and Wang, R.Y., Interactions of cholecystokinin octapeptide and dopamine on nucleus accumbens neurons, *Brain Res.*, 300, 161-166, 1984.

Whiteford, H.A., Stedman, T.J., Welham, J., Csernansky, J.G. and Pond, S.M., Placebo-controlled, double-blind study of the effects of proglumide in the treatment of schizophrenia, *J. Clin. Psychopharmacol.*, 12, 337-340, 1992.

Williams, R.G., Gayton, R.G., Zhu, W.-Y. and Dockray, D., Changes in brain cholecystokinin octapeptide following lesions of the medial forebrain bundle, *Brain Res.*, 213, 227-230, 1981.

Woodruff, G.N. and Hughes, J., Cholecystokinin antagonists, *Ann. Rev. Pharmacol. Toxicol.*, 31, 469-501, 1991.

Wu, T. and Wang, H.L., CCK-8 excites substantia nigra dopaminergic neurons by increasing a cationic conductance, *Neurosci. Lett.*, 170, 229-232, 1994.

Yim, C.C. and Mogenson, G.J., Electrophysiological evidence of modulatory interaction between dopamine and cholecystokinin in the nucleus accumbens, *Brain Res.*, 541, 12-20, 1991.

Záborszky, L., Alheid, G.F., Beinfeld, M.C., Eiden, L.E., Heimer, L. and Palkovits, M., Cholecystokinin innervation of the ventral striatum: a morphological and radioimmunological study, *Neuroscience*, 14, 427-453, 1985.

Zetler, G., Neuropharmacological profile of cholecystokinin-like peptides, *Ann. N.Y. Acad. Sci.*, 448, 448-469, 1985.

Zhang, J., Chiodo, L.A. and Freeman, A.S., Effects of the CCK-A receptor antagonist CR1409 on the activity of rat midbrain dopamine neurons, *Peptides*, 12, 339-343, 1991.

Zhang, J. and Freeman, A.S., Electrophysiological effects of CCK peptides on nondopaminergic neurons in rat substantia nigra pars reticulata, *Brain Res.*, 652, 154-156, 1994.

CHAPTER 5

THE MODULATION OF MIDBRAIN DOPAMINERGIC
SYSTEMS BY GABA

Judith R. Walters and Michele L. Pucak

I. HISTORICAL PERSPECTIVE

γ-Aminobutyric acid (GABA) was the first neurotransmitter to receive significant attention as a potential modulator of midbrain dopamine (DA) cell activity. This attention stemmed from evidence that GABA was a transmitter utilized by neurons innervating the substantia nigra. [Fahn and Côté, 1968; Kim et al., 1971; Okada et al., 1971; Precht and Yoshida, 1971] Initially, interest focused on the role of this transmitter as the agent mediating compensatory changes in DA cell activity in response to alterations in DA receptor stimulation. In 1963, a pivotal study by Carlsson and Lindqvist demonstrated that administration of the antipsychotic drug chlorpromazine induces a marked increase in the levels of DA metabolites in the striatum. To explain this effect, Carlsson and Lindqvist hypothesized that chlorpromazine was acting as a DA receptor blocker; the increase in DA metabolite levels was thought to be secondary to an increase in nigral DA cell firing which was mediated through a striatonigral feedback loop. [Carlsson and Lindqvist, 1963] This hypothesis made the neurotransmitter released by the striatonigral neurons a potentially critical link between effects of drugs acting at DA receptors and alterations in DA cell firing, DA synthesis and release. Experiments were designed to study the role of GABA in this process and the model was refined to include a third neuron, usually postulated to be an inhibitory striatal interneuron, in order to make the feedback loop work correctly and provide disinhibition of the DA neurons when striatal neurons postsynaptic to DA terminals were disinhibited by DA receptor blockade. [e.g. Bedard and Larochelle, 1973; Bunney and Aghajanian, 1976; Bunney et al., 1973; Guyenet et al., 1975; Kataoka et al., 1974; Ladinsky et al., 1976; Pérez de la Mora et al., 1975; 1976; Pérez de la Mora and Fuxe, 1977; Trabucchi et al., 1975]

The importance of GABA's role in relaying information about changes in the level of DA receptor stimulation back to the DA neuron became less clear in the mid 70's, when it was discovered that there were DA receptors (termed DA autoreceptors) on the terminals and cell bodies of the DA neurons themselves. [Aghajanian and Bunney, 1977; Kehr et al., 1972; Roth et al., 1973; Walters and Roth, 1974] Research showed that even in the absence of impulse flow in the DA neuron, DA agonists and antagonists could alter rates of DA synthesis and change DA metabolite levels in the striatum. [Kehr et al., 1972; Roth et al., 1973; Walters and Roth, 1974; 1976] These changes did, in fact, involve a feedback system

0-8493-4780-7/96/$0.00+$.50
© 1996 by CRC Press, Inc.

which adjusted DA synthesis in response to DA receptor stimulation, but it was a shortloop system, involving the action of DA on synthesis-modulating presynaptic DA autoreceptors in or near the synapse and was not dependent on changes in activity in the GABAergic striatonigral pathway. It also became evident that DA agonists could affect the firing rates of DA neurons directly, by acting at DA autoreceptors on the DA cell bodies. [Aghajanian and Bunney, 1977; Baring et al., 1980; Groves et al., 1975]

These considerations led to a revised view of GABA/DA interrelationships in which the hypothesized role of GABAergic inputs in mediating adjustments of DA cell firing rates in response to DAergic drugs became more limited. It became clear that changes in the firing rate of DA neurons induced by drugs which alter DA receptor stimulation did not necessarily provide insight into the function of the GABAergic input to the substantia nigra. Involvement of the GABAergic neurons in mediating the decreases in DA cell firing rates induced by DA releasing agents, such as amphetamine, and in mediating the increases in DA cell firing rates induced by DA receptor antagonists, such as haloperidol, remained a possibility. [Bunney and Aghajanian, 1978; Bunney and Grace, 1978; Hommer and Bunney, 1980; Iwatsubo and Clouet, 1977; Nakamura, et al., 1981; White and Wang, 1983] However, even these changes in DA cell firing could be attributed to other processes after it became appreciated that the DA neurons had not only autoreceptors, but also the apparent capability of releasing DA from their dendrites. [Björklund and Lindvall, 1975; Geffen et al., 1976; Nieoullon et al., 1977; Sladek and Parnavelas, 1975] If this dendritic release provided tonic stimulation for the local DA autoreceptors, then local effects might also be able to account for the actions of amphetamine and DA antagonists on DA cell firing rates [Paden et al., 1976; Pinnock, 1983; Wang, 1981a; b] and the role of the GABAergic striatonigral neurons could be reconsidered.

In the late 70's, focus began to shift from the dopaminocentric view wherein GABAergic input to the substantia nigra was considered to function primarily to provide feedback to the DA cells - to the idea that GABAergic modulation of nigral neurons served more as a feedforward system, relaying striatally processed information from the cortex to the substantia nigra as part of the basal ganglia output. [Collingridge and Davies, 1981; Di Chiara et al., 1978; Domesick, 1981; Garcia-Munoz et al, 1977; Marshall and Ungerstedt, 1977; Olianas et al., 1978] Similar issues began to be explored with respect to the role of GABA-containing projections to the DA neurons in the ventral tegmental area (VTA). [Arnt and Scheel-Krüger, 1979; German et al., 1980; Waddington and Cross, 1978; Walaas and Fonnum, 1980; Wang, 1981a; b; Yim and Mogenson, 1980] The GABAergic efferents from the striatum and accumbens continued to be viewed as potentially important sources of modulatory input to DA neurons in the midbrain, but the net effect of their influence on the DA neurons became more difficult to predict. Neurophysiological and anatomical studies supported the view that striatonigral GABAergic neurons not only connected directly with DA neurons, but also influenced them indirectly through the GABAergic cells in the substantia nigra pars reticulata. [Deniau, 1982; Grace and Bunney, 1979; Grofová et al., 1982;

MacNeil et al., 1978; Walters and Lakoski, 1978; Waszczak and Walters, 1979; 1980a; b; Waszczak et al., 1980] In addition, GABAergic inputs from the globus pallidus to both nigral neuronal systems provided additional complexity to the problem of sorting out GABA's role in the substantia nigra. [Hattori et al., 1975]

Since these early observations, both old and new techniques have continued to add to the picture of how GABAergic influences affect the activity of the midbrain DA neurons. The results of these studies will be reviewed below with the major emphasis given to how these studies have expanded understanding of the role of GABA with respect to the A9 neurons in the substantia nigra pars reticulata. Especially important has been the development of drugs selective for different DA and GABA receptor subtypes as well as the cloning and localization of many of these receptors. These approaches have led to substantial insight. Nevertheless, it should be pointed out that many challenging questions remain unanswered regarding the functional role of the GABAergic input to the DA neurons. It is clear, however, that GABA continues to live up to its early promise as a potentially important modulator of DA cell function.

II. GABA RECEPTORS

GABA can alter neuronal activity by interacting with either of two major receptor types, the $GABA_A$ and $GABA_B$ receptors. The $GABA_A$ receptor is an integral membrane protein which forms a chloride channel, and is also the site of action of benzodiazepines and barbiturates. [Pritchett et al., 1989b; Herb et al., 1992; Pritchett and Seeburg, 1990] This receptor is believed to be a heteropentameric glycoprotein composed of 5 heterogenous subunits (alpha 1-6, beta 1-3, gamma 1-3, delta, and rho 1-2), which have been described by a large number of cloning studies. [Bateson et al., 1991; Cutting et al., 1991; Herb et al., 1992; Khrestchatisky et al., 1989; Knoflach et al., 1991; Lasham et al., 1991; Levitan et al., 1988a; b; Lolait et al., 1989; Lüddens et al., 1990; Lüddens and Wisden, 1991; Malherbe et al., 1990; Nayeen et al., 1994; Pritchett et al., 1989a; b, Pritchett and Seeburg, 1990; Schofield et al., 1987; Shivers et al., 1989; Whiting et al., 1990; Wilson-Shaw et al., 1991; Wisden et al., 1991; Ymer et al., 1989a, Ymer et al., 1989b] These subunits are differentially expressed in various areas of the brain [Gambarana et al., 1991; Khrestchatisky et al., 1989; Levitan et al., 1988b; MacLennan et al., 1991; Persohn et al., 1992; Shivers et al., 1989; Wisden et al., 1992; Ymer et al., 1989a; b], and certain populations of neurons express specific assemblages of these subunits. [Fritschy et al., 1992; Gao et al., 1993] The differential expression of these subunits results in pharmacological heterogeneity of $GABA_A$ receptors. For example, the nature of the subunits present determine the degree of responsivity to benzodiazepines. [Herb et al., 1992; Pritchett et al., 1989b] The effects of benzodiazepines on $GABA_A$ receptors which include the alpha 1 subunit are pharmacologically distinguishable from those receptor assemblages with the alpha 2, 3, or 5 subunit. [Pritchett and Seeburg, 1990; Pritchett et al., 1989a; b] In addition, the efficacy of GABA may depend upon which types of subunit are present, as various combinations of subunits differ in their conductance and gating properties. [Angelotti and Macdonald, 1993; Herb et al., 1992; Khrestchatisky et al., 1989; Levitan et al., 1988b, Pritchett et al., 1989a; b]

The $GABA_B$ receptor is less well characterized than the $GABA_A$ receptor, since its cloning has lagged behind that of many neurotransmitter receptors. However, as a G-protein linked receptor [Holz et al., 1986], the $GABA_B$ receptor(s) is most likely to have a seven-membrane spanning region protein analogous to other G-protein-coupled receptors. Unlike the $GABA_A$ receptors, $GABA_B$ receptors are coupled to a variety of intracellular effector systems and induce changes in membrane physiology indirectly, resulting in increased potassium conductance and decreased calcium conductance. [Dunlap and Fischbach, 1981; Gähwiler and Brown, 1985; Knott et al., 1993; Mott and Lewis, 1994; Newberry and Nicoll, 1984] Pharmacological and functional heterogeneity of the $GABA_B$ receptor have been described. For example, presynaptically located $GABA_B$ receptors, which appear to attenuate the release of neurotransmitter, may be pharmacologically distinct from postsynaptic $GABA_B$ receptors, as well as from each other. [Bonanno and Raiteri, 1992; 1993a; b; Bowery et al., 1990; Calabresi et al., 1991; Dutar and Nicoll, 1988b] In addition, presynaptic attenuation of excitatory and inhibitory transmitter release by $GABA_B$ receptors may be mediated by different mechanisms. [Mott and Lewis, 1994] Resolution of the question of subtypes awaits the outcome of the cloning studies. It should also be pointed out that a third GABA receptor subtype, the $GABA_C$ receptor (also referred to as the $GABA_{NANB}$ (nonA, non-B) or $GABA_\rho$), has been described [for review, see Johnston, 1994] but not yet localized to the area of the midbrain DA neurons.

Athough both $GABA_A$ and $GABA_B$ receptors cause changes in cell physiology which can be classified as inhibitory [but see Staley et al., 1995], the fact that they act through different kinds of receptors and affect different ion conductances makes it possible for GABA to exert a range of effects on neuronal function. For example, because one is directly linked to an ion channel while the other operates via second messengers, the temporal characteristics of the effects produced by the two receptors are different; stimulation of $GABA_A$ receptors causes a short-latency, fast inhibitory postsynaptic potential (IPSP), whereas $GABA_B$ receptor stimulation results in a longer-latency, slow IPSP. [Dutar and Nicoll, 1988a] Furthermore, $GABA_B$ receptor stimulation appears to require higher concentrations of GABA than does the $GABA_A$ receptor stimulation and the slower time course of the $GABA_B$ response raises the possibility that these receptors may not necessarily be synaptically located. [for review, see Mott and Lewis, 1994] Thus, the two receptor subtypes and their distinct IPSP's may serve quite different functions. $GABA_A$ receptor-mediated IPSP's are capable of inducing large conductance changes and effectively reducing a neuron's ability to fire action potentials; $GABA_B$-mediated IPSP's more frequently induce relatively small conductance increases but large hyperpolarizations. $GABA_B$ receptor's postsynaptic effects appear to serve a more modulatory function; they are more readily overcome and have been shown more effective at inhibiting responses mediated by NMDA receptors than those mediated by non-NMDA receptors, because of the voltage dependence of the NMDA-mediated response. [Morrisett et al., 1991] $GABA_B$ receptors have also been shown able to play a role in priming burst firing activity through the reinactivation of voltage-inactivated conductances and the generation of low-threshold calcium spikes. [Crunelli and Leresche, 1991] Finally, presynaptic $GABA_B$ receptor mediated inhibitory effects on the release of glutamate, GABA and other neurotransmitters have been reported [for review, see Mott and Lewis, 1994], all of which makes it difficult to predict what the net effect of $GABA_B$ receptor stimulation will be at a given synapse. In addition, as discussed above, the $GABA_A$ receptor is clearly a heterogenous group of receptors with different pharmacology and gating properties, and this also may be the case for the $GABA_B$ receptor. Clearly, a complete understanding of the impact of GABAergic neurotransmission on the activity of midbrain DA neurons will require a comprehensive understanding of the specific types of GABA receptors localized on the DA neurons as well as on neuronal elements influencing them.

III. LOCALIZATION OF GABA RECEPTORS WITHIN THE SUBSTANTIA NIGRA

Various studies have demonstrated that nigral neurons possess GABA receptors. [Bowery et al., 1987; Lo et al., 1983; Nicholson et al., 1992; Pan et al., 1984; Reisine et al., 1979] Autoradiography experiments demonstrate high levels of $GABA_A$ receptors in the substantia nigra pars reticulata, with lower levels in the pars compacta. [Bowery et al., 1987; Nicholson et al., 1992] A potentially more important difference between the DA and non-DA neurons of the substantia nigra is that these two cell populations appear to express different subtypes of the $GABA_A$ receptor. Immunolabelling and *in situ* hybridization experiments are in general agreement that the $GABA_A$ receptors in the pars reticulata include primarily the alpha 1 subtype, along with beta 2 and 3 and gamma 1 and 2. In contrast, the alpha 3 and gamma 2 subtypes are the primary subtypes expressed in the zona compacta. [Araki et al., 1992; Fritschy et al., 1992; Malherbe et al, 1990; Nicholson et al., 1992; Persohn et al., 1992; Wisden et al., 1992; Zhang et al., 1991]

Autoradiographic studies have also indicated that $GABA_B$ receptors are present in moderate levels in the substantia nigra. Although the localization of this receptor subtype with respect to DA versus non-DA cell types in the substantia nigra is less characterized than that of $GABA_A$, as was observed for $GABA_A$ receptors, the levels of $GABA_B$ receptors are higher in the zona reticulata than in the zona compacta. [Bowery et al., 1987] Elucidating the pre- and postsynaptic distribution of $GABA_B$ receptors on the various cellular elements in the substantia nigra awaits cloning studies and the ability to identify and localize relevant mRNAs to the neuronal elements projecting into and out of this area.

Physiological and pharmacological techniques have also provided evidence supporting the presence of $GABA_A$ and $GABA_B$ receptors on DA neurons, as discussed below. DAergic as well as non-DAergic neurons in the substantia nigra and VTA respond to the application of GABA and GABA agonists selective for

GABA$_A$ and GABA$_B$ receptor subtypes [e.g. Collingridge and Davies, 1981; Grace and Bunney, 1979; Jiang et al., 1993; Lacey et al., 1988; Olpe et al., 1977; Pinnock, 1984; Waszczak et al., 1980a; b] and there is some evidence for the presence of presynaptic GABA$_A$ and GABA$_B$ receptors on terminals of nigral afferents. [Arbilla et al., 1979; Giralt et al., 1990; Häusser and Yung, 1994; Nicholson et al., 1992]

IV. SOURCE OF GABAERGIC TERMINALS IN THE SUBSTANTIA NIGRA

The substantia nigra contains high levels of GAD and a large number of GABAergic terminals [Okada et al., 1971; Ribak et al., 1976; Tappaz et al., 1976]. As discussed above, this information led to the early recognition of the potential importance of GABA in the regulation of DA neurons. Indeed, the critical importance of GABA was suggested by the observations that 1) most terminals in the nigra contain GAD and 2) nigral dendrites are often surrounded by GAD-positive terminals. [Bolam and Smith, 1990; Mendez et al., 1993; Ribak et al., 1976] These terminals form primarily axodendritic but also axosomatic synapses with neural elements in both the pars compacta and the pars reticulata [Smith and Bolam, 1989; Smith and Bolam, 1990], and have been shown to form synapses onto both tyrosine hydroxylase-positive and negative elements. [Mendez et al., 1993; Smith and Bolam, 1990; Van Den Pol et al., 1985] The GABAergic terminals arise from several forebrain nuclei, including the striatum, globus pallidus, and nucleus accumbens, as well as from GABAergic neurons which are intrinsic to the substantia nigra.

A. INTRINSIC SOURCES
Although the majority of cells in the pars compacta are DAergic, neurons which are immunoreactive for GAD have been described in both the pars reticulata and the pars compacta. In particular, pars reticulata neurons which project to the thalamus, superior colliculus, and other caudal nuclei have been suggested to utilize GABA as a neurotransmitter. [DiChiara et al., 1979; Kilpatrick et al., 1980; MacLeod et al., 1980; Ueki et al., 1977; Vincent et al., 1978] Neurons of this type are medium-sized, with three to five varicose dendrites oriented in the medio-lateral and anterio-posterior directions. [Fallon and Loughlin, 1985; Grofova et al., 1982] The axons of these neurons form one to five collaterals which ramify within the parent field as well as nearby but outside it, in both the pars reticulata and the pars compacta. [Deniau et al., 1982; Grofova et al., 1982] Terminals of these neurons have been observed to synapse onto DAergic neurons [Damlama et al., 1993]; thus, these GABAergic neurons are likely to modulate DA cell activity. In addition, small neurons believed to be interneurons are present in the substantia nigra, within both the pars reticulata and the pars compacta. [Francois et al., 1979; Schwynn and Fox, 1972] These cells have approximately three short dendrites and their axons have been observed to run along the dendrites of medium-sized neurons in the substantia nigra. On the basis of physiological and pharmacological

data, these have been suggested to be inhibitory and thus may contain GABA [Grace and Bunney, 1979], but this has not been directly demonstrated. Either of these types of neurons could account for the observation that after hemisections between the substantia nigra and the forebrain, some GAD-immunoreactive boutons are spared; these form synapses onto both GAD-positive and GAD-negative, presumably DAergic, neurons in the substantia nigra. [Nitsch and Riesenberg, 1988]

B. EXTRINSIC SOURCES

Many studies have demonstrated that a major source of GABAergic input to the substantia nigra arises from the striatum. [Bolam and Smith, 1990; Brownstein et al., 1977; Fonnum et al., 1974; Grofova, 1975; Jessell et al., 1978; Kim et al., 1971; Ribak et al., 1980] In this projection, GABA is colocalized with substance P, dynorphin, and, in a small number of terminals, with enkephalin. [Reiner and Anderson, 1990] The striato-nigral projection terminates throughout the substantia nigra, although, from a quantitative point of view, its primary target is the pars reticulata [Grofova and Rinvik, 1970; Hattori et al., 1975; Smith and Bolam, 1991; Somogyi et al., 1981; Wasseff et al., 1981], which contains mostly non-DA cells, as well as the dendrites of those DA cells which are located in the ventral pars compacta. [Fallon and Loughlin, 1985; Grace and Bunney, 1983b, Grace and Onn, 1989; Gulley and Wood, 1972; Preston et al., 1981; Tepper et al., 1987] The terminals of striato-nigral afferents are small and numerous, with few mitochondria. They synapse primarily on small, distal dendrites in the pars reticulata [Grofova and Rinvik, 1970; Hattori et al., 1975; Smith and Bolam, 1991; Somogyi et al., 1981; Wasseff et al., 1981] and have been shown to form synapses onto DA and onto non-DA neuronal elements. [Bolam and Smith, 1990; DiChiara et al., 1979; Nitsch and Riesenberg, 1988; Somogyi et al., 1979; Von Krosigk et al., 1992]

There is clearly direct regulation of DA neurons by cells in the striatum, as terminals of the GABAergic striato-nigral pathway have been demonstrated to form synapses onto identified DA neurons. [Bolam and Smith, 1990] Nonetheless, the prevalence of terminals in the pars reticulata rather than the pars compacta has led to the suggestion that the striatum exerts a potent regulatory effect on the non-DA pars reticulata cells, whereas its influence over DA neurons is mediated primarily indirectly, via these non-DA reticulata cells. [Grace and Bunney, 1979] The extent of the direct versus indirect control of any particular DA neuron by GABAergic inputs may depend on the characteristics of the cell in question. For example, DA cells located in the ventral tier of the pars compacta, with dendrites which extend into the pars reticulata, may be regulated by direct striatal input to a greater extent than are the DA neurons in the dorsal tier, whose dendrites remain within the pars compacta. Furthermore, DA cells may be differentially regulated by striatonigral afferents originating in different compartments of the striatum; the striatal patches project primarily to the pars compacta, whereas the striatal matrix projects primarily to the pars reticulata. [Gerfen, 1985] Thus, DA neurons located in the dorsal pars compacta may be regulated solely by the patches, while ventrally-located DA neurons, whose dendrites are in the pars compacta but also enter the pars reticulata, may be modulated by both types of striatal compartments.

The substantia nigra also receives GABAergic input from the globus pallidus. [Grofova, 1975; Hattori et al., 1973; Smith and Bolam, 1989; Smith and Bolam, 1991] Like the striatonigral projection, this projection appears to terminate primarily in the pars reticulata. [Smith and Bolam, 1990; Smith and Bolam, 1991; but see Hattori et al., 1975] Boutons of the pallido-nigral projection are larger than those of the striato-nigral projection and form synapses primarily onto somata and proximal dendrites of non-DA output neurons of the pars reticulata, with a smaller component forming synapses on the somata and dendrites of DA neurons. [Smith and Bolam, 1990; 1991; Von Krosigk et al., 1992] There is also a relatively small projection from the nucleus accumbens, which terminates on nigrostriatal DAergic neurons. [Somogyi et al., 1981]

V. EFFECTS OF GABA ON DA CELL PHYSIOLOGY

The high density of GABAergic input as well as the existence of GABAergic receptors within the substantia nigra suggests that this neurotransmitter plays an important role in modulating the activity of nigral DA neurons. Indeed, the physiological and pharmacological studies discussed below have supported the significance of this role and the potential for GABA to exert a significant influence on the firing pattern as well as the rate of the DA neurons. On the other hand, these studies have also shown that DA neurons are not as sensitive to the inhibitory effects of applied GABA as are, for example, the neighboring neurons in the substantia nigra pars reticulata. Most notably, however, these studies are impressive in their efforts to sort out the individual effects exerted by interactions of GABA with the many sites through which this transmitter can act to affect DA cell function. DA neurons are regulated by GABAergic inputs from within the substantia nigra and from extrinsic sources; in addition, neurotransmitter release from these GABAergic inputs and from other afferents may be modulated by presynaptic GABA receptors. As these mechanisms have been investigated, results are consistent with the proposal that DA neurons are modulated by GABAergic neurons which are intrinsic to the substantia nigra as well as by GABAergic afferents, and this latter form of regulation may be mediated directly as well as indirectly, via the intrinsic GABAergic neurons. Similar modulation of A10 DA neurons via intrinsic as well as extrinsic afferents appears to exist in the VTA.

A. *IN VITRO* EFFECTS OF GABA AND GABAERGIC DRUGS
1. Postsynaptic Effects
Studies utilizing *in vitro* slice preparations have been useful in elucidating the direct effects of GABA on DA neurons. Application of GABA to DA neurons in the substantia nigra or the VTA causes hyperpolarization, decreased input resistance, and outward current. [Lacey et al., 1988; 1989; Mereu et al., 1992; Pinnock, 1984] These inhibitory effects can be mediated by either receptor

subtype: the $GABA_A$ receptor agonist muscimol causes decreased input resistance and outward current, presumably secondary to increased chloride conductance [Lacey et al., 1988; Pinnock, 1984], and the $GABA_B$ receptor agonist baclofen also causes hyperpolarization, decreased input resistance, and an outward current in DA neurons. [Lacey et al., 1988; 1989; Pinnock, 1984] Effects of baclofen have also been observed when drugs are applied to neurons in dissociated cell preparations [Roeper et al., 1990], and are not altered by addition of tetrodotoxin (TTX). [Lacey et al., 1988] Like DA cells, non-DA nigral neurons are also inhibited by application of drugs which interact with either $GABA_A$ and $GABA_B$ receptors in slice preparations. [Lacey et al., 1989; Mereu et al., 1990; Rick and Lacey, 1994]

a. $GABA_A$ Receptor Effects and Benzodiazepines

Consistent with reports that the pars reticulata and pars compacta may contain different $GABA_A$ subunits, neurons located in these regions differ in their sensitivity to benzodiazepines which modulate the function of the $GABA_A$ receptor. Some drugs, such as diazepam, have similar effects on both types of neurons, whereas others, such as flunitrazepam and zolpidem, cause a greater enhancement of the peak inhibitory postsynaptic current (IPSC) amplitude in pars reticulata than in DA neurons. [Mereu et al., 1992] The different responses of the $GABA_A$ receptors to certain benzodiazepines appear related to the identity of the receptor's alpha subunit, and the enhanced response of the substantia nigra pars reticula neurons to zolpidem and flunitrazepam may reflect the higher concentration of the alpha 1 subunit in the pars reticulata. [Pritchett and Seeburgh, 1990; Pritchett et al., 1989a]

b. $GABA_B$ Receptor Effects on Second Messengers and Neuronal Excitation

In vitro studies have shown that $GABA_B$-mediated effects on DA cells are caused by facilitation of a potassium conductance, and are altered by inclusion of GTP-gamma-S in the recording electrode. [Lacey et al., 1988] Although the precise identity of the potassium currents which mediate the effects of $GABA_B$ receptor stimulation have not been described, these in vitro intracellular recording studies suggest that the $GABA_B$ receptor may facilitate the same potassium conductance which is facilitated by stimulation of the D_2 receptor [Lacey et al., 1988], a finding which may have important implications for GABA-DA interactions at the single-cell level. Both these receptors are known to be G-protein linked, and thus it has been suggested that they may share a second messenger system. In addition, baclofen has also been reported to decrease a hyperpolarization-induced cation current, and this effect is abolished by pretreatment with GTP-gamma-S. However, this effect probably does not mediate the decreases in firing rate caused by baclofen, because the inhibition was small and because this current is largely inactivated at normal membrane potentials. [Jiang et al., 1993]

Although the hyperpolarizing effect of GABA is normally viewed as having the consequence of inhibiting the firing of action potentials, it has been pointed out that the opposite effect may also be possible. In thalamocortical cells, for example, long-duration hyperpolarizing events such as those activated by the GABA$_B$ receptor have been shown to reactivate voltage-inactivated conductances, regulate low-threshold calcium spikes and induce bursts of action potentials. [for review, see Crunelli and Leresche, 1991; Mott and Lewis, 1994] Häusser and Yung have discussed the possibility of a similar phenomenon occurring with respect to DA cell activity. [Häusser and Yung, 1994] It appears possible that under some conditions, GABA$_B$-mediated long-latency IPSP's could cause a rebound activation of intrinsic currents in a quiescent DA cell, leading to a low threshold spike and activating the repetitive oscillations in membrane potential which have been described as underlying the endogenous spontaneous activity observed in DA cells *in vivo* and *in vitro*. [Grace and Bunney, 1984; Llinas et al., 1984] If so, the direct effect of GABA, depending on the state of the DA cell and the presence of GABA$_B$ receptors at the synapse, could occasionally be excitatory, adding a further level of complexity to the potential effects of GABA on DA cell activity.

c. Differential Effects of GABAergic Input on GABA$_A$ and GABA$_B$ Receptors

In addition to evoked GABA-mediated effects, spontaneous GABAergic IPSP's and IPSC's are observed in both DA and non-DA nigral neurons recorded in nigral slices. [Hajós and Greenfield, 1993; Häusser and Yung, 1994; Johnson and North, 1992a; b; Mereu et al., 1992; Pessia et al., 1994; Rick and Lacey, 1994; Sugita et al., 1992] These appear to be mediated by GABA$_A$ receptors, because they are blocked by bicuculline. In fact, bicuculline has been reported to cause marked increases in rate in substantia nigra pars reticulata neurons in slices, and modest increases in some DA cells. [Pinnock, 1984; Rick and Lacey, 1994] These observations provide important insight into relationships between striatonigral and intranigral GABAergic neurons and DA cells. Similar interrelationships appear to exist between these neurons' counterparts in the VTA. It is convincingly argued that the GABA$_A$-mediated events observed in both DAergic and non-DAergic neurons in these *in vitro* studies arise from activity within GABAergic neurons which are intact within the slice. This idea is consistent with the observation that the frequency of these events is decreased by application of TTX. [Häusser and Yung, 1994; Johnson and North, 1992a; b] Although GABA$_B$-mediated IPSP's can be induced by stimulation of the slice, spontaneous GABA$_B$-mediated events are not observed, presumably because these receptors are innervated by neurons whose cell bodies lie outside the substantia nigra or VTA. Thus, these studies suggest that whereas GABA$_A$ receptors on DA neurons are innervated by GABAergic nigral neurons, the GABA$_B$ receptors are innervated by GABAergic afferents originating outside the substantia nigra, presumably in the globus pallidus and/or striatum. [Häusser and Yung, 1994; Johnson and North, 1992a; b; Sugita et al., 1992] However, it should be noted that stimulation of nigral afferents from the striatum causes monosynaptic inhibition of DA neurons which appears to be mediated by stimulation of GABA$_A$ receptors. [Collingridge and Davies, 1981; Grace and Bunney, 1985, Precht and Yoshida, 1971] Taken together, these observations raise the possibility that the two GABA receptor subtypes are differentially located with respect to the GABAergic innervation of the substantia nigra neurons in such a way that the pallidonigral neurons, whose terminals have been localized to the proximal areas of the DA neurons [Smith and Bolam, 1990; Smith and Bolam, 1991; Von Krosigk et al., 1992], may be preferentially responsible for the GABA$_B$-mediated IPSP's induced by stimulation of the slice preparation. It may be worth noting that the pallidal neurons are more tonically active *in vivo* than the striatonigral neurons. Because of the different mechanisms underlying the IPSP's induced by the two GABA receptor subtypes - the rapid receptor-mediated change in Cl$^-$ conductance versus the slower, second-messenger-mediated change in K$^+$ conductance - the possibility that the different GABA receptor subtypes may be differentially distributed with respect to the major afferents to the DA neurons could have consequences with respect to the impact of these individual afferents on short- and long-term (e.g. immediate early gene-mediated) aspects of DAergic function.

2. Presynaptic Effects on GABA Release

In addition to providing insight into the source of the GABAergic influences on DA neurons, these *in vitro* slice studies have also called attention to a) the ability of GABA to act presynaptically to modulate its own release, and the release of other transmitters and b) the ability of other transmitters to act presynaptically to modulate the release of GABA from local GABA afferents in the substantia nigra and VTA.

a. Presynaptic GABA Receptors

The modulatory role of presynaptic GABA receptors on neurotransmitter release has been investigated in many brain regions. In particular, GABA$_B$ presynaptic receptors have been studied with respect to their apparent ability to suppress the release of excitatory neurotransmitter and to regulate the release of GABA. [Mott and Lewis, 1994] Studies in substantia nigra slices have also explored the role of presynaptic GABA$_A$ and GABA$_B$ receptors in modulating the release of GABA onto DA neurons. Examination of the effect of selective GABA antagonists on the response of DA neurons to paired pulse stimulation has provided data to support the view that presynaptic GABA$_B$ receptors play a role in modulating GABA release in the nigra. [Giralt et al., 1990; Häusser and Yung, 1994] Evidence for a presynaptic role for GABA$_A$ receptors was not found in these neurophysiological studies although there is anatomical evidence for presynaptically located GABA$_A$ receptors. [Nicholson et al., 1992]

b. Local Effects of Non-GABA Receptors on GABA Release

Any neurotransmitter receptor expressed by the GABAergic afferents to the DA cells has the potential to modulate DAergic function indirectly, by affecting GABA release. Most commonly, this is thought of as being mediated through receptors located on the dendrites and cell bodies which regulate the firing rate of the GABAergic afferent. However, an additional mechanism involves presynaptic effects on synaptic release of transmitter and these will be discussed here. In fact, a number of different neurotransmitter receptors appear to be able to mediate changes in the activity of DA cells in slice preparations by altering the release of GABA from local afferents. The best studied is the D_1 DA receptor. Anatomical studies have demonstrated that D_1 receptors are highly concentrated in the substantia nigra and are located on the terminals of the striatonigral GABAergic neurons. [Porceddu et al., 1986; Altar and Hauser, 1987] Evidence from slice studies and *in vivo* experiments indicate that these receptors can mediate an increase in GABA release. [Aceves et al., 1992; Cameron and Williams, 1993; Floran et al., 1990; Reubi et al., 1977; Timmerman et al., 1991] and there is considerable interest in the role of these presynaptic D_1 receptors in mediating functionally relevant behavioral effects of DA and DA agonists. [LaHoste and Marshall, 1990; Robertson and Robertson, 1987]

Because of its relevance for the mechanism of action of cocaine, there is also significant interest in the possibility that serotonin (5HT) may affect the activity of DA neurons by exerting a local effect on GABA release in the VTA. Studies have indicated that the response of the VTA DA neurons to 5HT receptor stimulation is mediated, in part, by effects of 5HT on GABAergic neurons within the slice. [Pessia et al., 1994] In addition, consistent with the proposed differential innervation of $GABA_A$ and $GABA_B$ receptors, the release of GABA onto $GABA_B$ but not $GABA_A$ receptors on VTA DA neurons appears to be modulated by stimulation of presynaptic $5\text{-}HT_{1D}$ receptors on nigral afferents. [Cameron and Williams, 1994; Johnson et al., 1992a] Also of interest because of relevance to drugs of abuse is evidence that local GABAergic innervation of DA neurons plays a role in mediating the excitatory effects of opioids on DA cell activity in the VTA: μ-receptor-mediated hyperpolarization of local GABAergic neurons has been shown associated with a reduction in the frequency of spontaneously occurring IPSP's in DA cells. [Lacey et al., 1989; Johnson and North, 1992a] In addition, stimulation of presynaptic adenosine receptors in the VTA affects GABA release in slice preparations [Wu et al., 1995] and modulation of GABA release from strionigral neurons by muscarinic receptors has been demonstrated. [Kayadjanian et al., 1994] The extent to which any or all of these processes is - or are - functionally relevant to the endogenous effects of these transmitters or the pharmacological effects of drugs on DA system function remains to be established. It is clear, however, that the potential for these mechanism to affect GABA release in the nigra and VTA certainly adds to the complexity of local wiring diagrams and to the difficulty of sorting out the roles of the various GABAergic projections to the midbrain DA neurons.

B. *IN VIVO* EFFECTS OF GABAERGIC DRUGS

In vivo studies have provided a perspective on the importance of GABA-mediated inhibitory systems to normal CNS function. This importance is attested to by consideration of some of the pharmacological effects of GABAergic drugs. For example, systemic administration of GABAergic antagonists such as bicuculline and picrotoxin induces convulsions; the $GABA_A$ agonist muscimol has hallucinogenic effects; and benzodiazepines and barbiturates, modulators of the

GABA$_A$ receptor/channel complex, are anxiolytics and anesthetics, respectively. GABA$_B$ agonists induce a form of anesthesia with loss of righting reflex and have been considered a model for absence seizures. While it is not clear that any of these effects involve GABA/DA system interactions specifically, many of these drugs have significant effects on DA cell activity when given systemically, and the substantia nigra has been implicated as a potentially important site of action for anticonvulsant drugs and benzodiazepines. Moreover, in spite of the complexity associated with the multiplicity of sites at which GABA can act to affect DA cell activity, *in vivo* studies have presented some very useful insights into the nature of the effects of GABAergic systems on DAergic function.

1. Stimulation Studies

Stimulation of regions sending GABAergic afferents to the nigra has provided insight into how DA cells are regulated by such inputs. Early studies demonstrated that stimulation of the striatum evokes a monosynaptic inhibitory postsynaptic potential (IPSP) in nigral neurons. [Yoshida and Precht, 1971] These inhibitory effects appear to be mediated by GABA$_A$ergic neurotransmission, because the associated field potentials are blocked by systemic administration of picrotoxin. [Collingridge and Davies, 1981; Precht and Yoshida, 1971] Subsequent identification of the DA neuron extracellular waveform [Bunney et al., 1973; Grace and Bunney, 1983a] permitted investigation of the striatal regulation of DA and non-DA neurons in the substantia nigra. Stimulation of the striatum elicits a short-latency IPSP in DA neurons which is chloride-mediated [Collingridge and Davies, 1981; Grace and Bunney, 1985], and thus occurs via stimulation of GABA$_A$ receptors. Such studies have therefore clearly demonstrated that stimulation of the striatum can cause direct GABA$_A$-mediated inhibition of both DA and non-DA neurons in the substantia nigra. [Grace and Bunney, 1985; Precht and Yoshida, 1971; Collingridge and Davies, 1981]

Although stimulation of the striatum causes a direct inhibition of DA neurons, it has been proposed that this is not the primary effect of the striatum on DA cells. As described above, *in vitro* studies have provided evidence for the potential for GABAergic inputs to induce both direct and indirect effects on DA cell activity. Indirect effects of striatal output could be mediated by both pallidal and nigral GABAergic input to the DA neurons. The idea that intervening, non-DA neurons within the substantia nigra play a prominent role in mediating the effect of the striatal neurons on DA cell activity was initially supported by a series of *in vivo* studies. Striatal stimulation evokes IPSPs in non-DA cells as well as DA cells in the nigra; these non-DA cell IPSPs are larger in amplitude and longer in duration than those evoked in DA cells. In addition, the IPSPs of DA cells are followed by a rebound depolarization which occurs at a time when the non-DA neuron is still inhibited. [Grace and Bunney, 1985] Furthermore, lower-intensity stimulation of the striatum actually causes an increase in the firing rate of DA cells which is temporally coincident with a decreased pars reticulata cell firing rate. [Grace and Bunney, 1985] Finally, chemical stimulation of striatal cells by infusion of kainic acid into the striatum also causes short-term excitation of DA cell activity. [Braszko et al., 1981]

These observations led to the suggestion that the striatum can alter DA neuron activity by causing either 1) a direct inhibition or 2) an indirect disinhibition mediated by inhibition of an intervening inhibitory non-DA cell. Indeed, this latter arrangement is consistent with early hypotheses about the role of the striatonigral pathway as a negative feedback loop, with the predominant effect of the striatum on DA cell activity being disinhibition providing for an increase in the activity of DA neurons during situations of low striatal DA. This proposal is consistent with the observations that infusion of the DA antagonist haloperidol into the striatum results in an increase in DA cell firing rate [Iwatsubo and Clouet, 1977], whereas striatal infusion of apomorphine causes a decreased DA cell firing rate. [Scarnati et al., 1980]

2. Systemic Administration
a. GABA_A Agonists and Antagonists

The model outlined above requires that the non-DA neurons in the substantia nigra be more responsive than the DA neurons to the direct effects of striatal stimulation. When this possibility was examined by comparing the response of DA and non-DA neurons to the direct application of GABA, it was found that the direct application of GABA *in vivo* by iontophoresis caused inhibition of both DA and non-DA nigral neurons. However, the inhibition observed in DA neurons was surprisingly modest, especially in comparison to the striking inhibition of non-DA neurons in the pars reticulata. This was observed whether GABA itself or a GABA agonist such as muscimol or 4,5,6,7-tetrahydroisoxazolo-[5,4-c]-pyridin-3-ol (THIP) are applied. [Waszczak et al., 1981; Waszczak and Walters, 1980; Grace and Bunney, 1979; Waszczak et al., 1980] It could be argued that this apparent difference in sensitivity might be an artifact of the manner in which the drug is delivered; for example, the GABA receptors of the non-DA neurons might be located closer to the cell bodies, and therefore be more accessible to iontophoretic drug application. However, pressure ejection studies suggest that the GABA receptors on DA neurons are not located distally. [Akaoka et al., 1992]

The effects of systemically administered GABAergic drugs on DA and non-DA cell activity also appeared consistent with a differential sensitivity of these cell populations to GABA_A receptor stimulation. When muscimol first became available for pharmacological studies, it presented an opportunity to examine the systemic effects of a drug which was relatively selective for GABA_A receptors and which was modestly capable of crossing the blood/brain barrier after systemic administration. Investigation of its systemic effects on the activity of DA cells showed that this drug actually caused an increase in DA neuronal firing rate, and this increase overlapped with a decrease in firing rate observed in pars reticulata neurons. [Engberg et al., 1993; Grace and Bunney, 1979; Walters and Lakoski, 1978; Waszczak and Walters, 1979; Waszczak et al., 1980a; b] In addition, the effects of systemically administered benzodiazepines on DA and substantia nigra pars reticulata cell activity produced similar effects. These drugs, which enhance the effects of GABA at $GABA_A$ receptors, acted a lot like muscimol with respect to nigral cell activity; they reduced the firing rates of the pars reticulata cells and produced a modest increase in the activity of the pars compacta dopamine cells. [Mereu and Biggio, 1983; Ross et al., 1982; Waszczak, 1983; Waszczak et al., 1986; Waszczak and Walters, unpublished observations]

These observations further supported the idea that the pars reticulata neurons exert a tonic inhibitory influence over DA neuron activity, and that suppression of the reticulata cell activity by stimulation of GABA receptors results in a disinhibition of the DA neurons. [Grace and Bunney, 1979; Waszczak, et al., 1980a; b] This hypothesis implies that the non-DA neurons have a greater sensitivity to GABAergic drugs than do the DA neurons; however, it also implies that the DA neurons themselves are somewhat responsive to GABA, since the reduction in tonic $GABA_A$-mediated inhibition arising from the pars reticulata cells is proposed to account for the increased firing rates observed in the DA cells following GABA agonist administration. Furthermore, the model requires that the reduction in pars reticulata GABAergic input to the DA neurons is not compensated for by the presence of muscimol acting at the same receptors. A number of possibilities may be relevant to explaining this seemingly inconsistent set of conditions; it may be that pharmacological differences in the subtypes of $GABA_A$ receptors on the DA versus the non-DA neurons result in a higher sensitivity of the non-DA neurons to particular GABAergic drugs. It is possible that presynaptic GABA receptors are playing a role in the dynamic equilibrium which produces the increase in DA cell activity after the systemic administration of a $GABA_A$ agonist. In addition, it is also possible that the systemic effects of muscimol may result in decreases in the activity of GABAergic neurons providing input to the $GABA_B$ receptors on the DA cells, reducing tonic GABA-mediated inhibition to sites insensitive to muscimol (i.e. $GABA_B$ receptors). However, it should be pointed out in this regard that kainic acid-induced lesion of the striatum did not reduce the response of the DA neurons to systemically administered muscimol, nor did this lesion reduce the ability of systemically administered picrotoxin to increase the activity of the pars reticulata neurons. [Waszczak et al., 1981] These experiments indicate that the effects of these systemically administered GABA agonists or antagonists on activity of the substantia nigra neurons are not primarily dependent upon inhibiting or blocking the effects of the strionigral afferents, and they argue for the importance of locally mediated effects of GABAergic drugs on DA neurons, presumably mediated through inhibiting or blocking the effects of the GABAergic neurons in the substantia nigra pars reticulata.

Additional studies also are consistent with this interpretation. For example, iontophoretic application of GABA into the pars reticulata causes an increase in DA neuron firing rate, whereas glutamate causes a decrease. [Grace and Bunney, 1979] Furthermore, DA neurons respond to footpinch with an inhibition-excitation sequence, whereas the opposite response is observed for many non-DA reticulata cells. [Grace and Bunney, 1979; Grace and Bunney, 1980] Also consistent with indications that there is a tonic inhibitory GABAergic input onto the DA cells is the fact that GABA$_A$ antagonists picrotoxin and bicuculline, administered systemically, cause increases in the firing rates of DA neurons [Grace and Bunney, 1985; Waszczak et al., 1980; 1981] These increases are modest, however, relative to those observed in the substantia nigra pars reticulata and the globus pallidus [Waszczak et al., 1980; 1981] and, as with all systemically administered drugs, it is possible that these changes in firing rate could involve indirect effects; disinhibition of excitatory input, for example, as well as direct blockade of inhibitory input. Direct application of bicuculline also causes increases in DA neuronal firing rate, however, and an increase in the number of cells firing in burst mode. [Tepper et al., 1995]

One question which has been raised during the course of these investigations is whether the pars reticulata neurons involved in regulating dopamine cell activity are interneurons, in the sense that they constitute a set of cells which send no projections to sites outside the substantia nigra, or whether they belong to, and/or include, the major population of reticulata neurons which project to the thalamus and superior colliculus but which appear also to have recurrent collaterals terminating within the substantia nigra. [Francois et al., 1979; Grace and Bunney, 1979; Schwynn and Fox, 1972] Evidence from a recent *in vivo* study has demonstrated inhibitory effects on DA cells of stimulation in the thalamus and superior colliculus.   This inhibition is blocked by the GABA$_A$ antagonist bicuculline, but not by the GABA$_B$ antagonist, 2-hydroxysaclofen. [Tepper et al., 1995] These results indicate that antidromic activation of reticulata neurons which project to the thalamus results in inhibition of DA neurons, presumably via activation of axon collaterals. This study demonstrates that at least a major population of the pars reticulata neurons mediating these effects on DA cell activity are output neurons which also project to the thalamus and superior colliculus (Anderson et al., 1993; Tepper et al., 1995].

b. GABA$_B$ Agonists and Antagonists

In contrast to the excitatory effects of GABA$_A$ agonists on DA cell firing, systemic administration of GABA$_B$ agonists can bring about a compete cessation of extracellularly recorded activity in the A9 and A10 DA neurons. Similar effects are observed with systemic administration of the GABA$_B$ agonist, baclofen [Grace and Bunney, 1980; Mereu et al., 1986], and with two other drugs identified as probable GABA$_B$ agonists, gamma-hydroxybuyric acid (GHB) and 3-amino-1-hydroxy-pyrrolid- 2-one (HA-966). [Engberg and Nissbrandt, 1993; Engberg et al., 1993; McMillen et al., 1992; Shepard and Lehmann, 1992; Shepard et al., 1995; Singh et al., 1990; Waldmeier, 1991; Walters et al., l972; 1973; Walters and Roth, 1974; Xie and Smart, 1992] It has been appreciated for many years that the latter two drugs inhibit DA cell firing, but it is only relatively recently that these effects have been attributed to stimulation of central GABA$_B$ receptors.

GHB and its precursor, gamma-butyrolactone (GBL), which is metabolized to GHB *in vivo* [Roth and Giarman, 1966], have a long history with respect to effects on the DA system. This compound initially received attention for its ability to produce a relatively unique anesthetic state, leaving the subject capable of autonomic response to stress (Rubin and Giarman, 1947; Laborit et al., 1960). Subsequently, it was found that systemic administration of GHB (or GBL) induced marked increases in DA levels in the striatum [Gessa et al., 1966] and it was thought that this drug might be exerting an excitatory effect on DA cell firing rates. The first DA cell recordings were initiated to explore this question [Walters et al., 1972] and it became apparent that GHB did not produce an increase in the activity of substantia nigra DA neurons; rather it produced a decrease. [Walters et al., 1972; 1973; Walters and Roth, 1974] Systemic administration of GBL also induced a decrease in the activity of cells in the substantia nigra pars reticulata and in other brain regions such as the globus pallidus, while 5-HT-containing neurons in the dorsal raphe and noradrenergic neurons in the locus coeruleus were not inhibited by this drug. [Diana et al., 1993; Walters et al., 1972; Walters and Roth, 1977] GHB became widely used to provide a model preparation helpful for screening effects of drugs on DA autoreceptors because, in the absence of neuronal impulse flow, DA agonist and antagonist effects on DA synthesis could be attributed to actions on the terminal DA autoreceptor. [Walters and Roth, 1976] Only after selective $GABA_B$ antagonists became available was it clear that this drug acts by stimulating $GABA_B$ receptors. [Engberg and Nissbrandt, 1993; Engberg et al., 1993; Xie and Smart, 1992]

It is now evident that the *in vitro* and *in vivo* effects of $GABA_B$ receptor stimulation on DA neurons are quite consistent with each other. Both approaches indicate that stimulation of $GABA_B$ receptors brings about a reduction in the activity of DA cell firing. The $GABA_B$ agonists also induce a marked regularization in DA neuronal firing pattern. [Engberg and Nissbrandt, 1993; Engberg et al., 1993; Grace and Bunney, 1980; McMillen et al., 1992; Shepard and Lehmann, 1992; Shepard et al., 1995] This ability to reduce rate and induce a pacemaker-like mode of firing is shared by D2 receptor agonists. It is interesting to note that pretreatment with the D2 antagonist, haloperidol, has been reported to enhance the sensitivity of DA cells to baclofen [Grace and Bunney, 1980; see also Mereu et al., 1986] raising questions about the possibility of interactions between intracellular mechanisms triggered by D2 and $GABA_B$ receptors. It should also be noted that after systemic administration of a $GABA_B$ agonist, DA cells typically show a modest increase in rate and bursting before they begin to slow down, and, over a certain range of low doses, baclofen, GHB or HA-966 can actually produce a sustained increase in DA cell firing. [Diana et al., 1991] This suggests that $GABA_B$ agonists, like $GABA_A$ agonists, can also induce disinhibition of DA cell firing, perhaps by inhibiting other inhibitory inputs more potently. Substantia nigra pars reticulata cells do not appear to be candidates for mediating this disinhibition, however. [Diana et al., 1993] Gessa and coworkers have suggested that this effect might have therapeutic potential in the treatment of alcoholism. These researchers have shown that ethanol also exerts excitatory effects on DA cell firing rates over a certain dose range [Mereu et al., 1984]; and in a clinical trial, treatment with GHB appeared to decrease craving and alcohol consumption over a 3 month period [Gallimberti et al., 1992; see also Cash, 1994].

These considerations lead to the interesting question of whether there is an endogenously significant action of GABA on the GABA$_B$ receptors mediating DA cell inhibition, and whether input to these receptors is tonically or phasically activated under normal or unusual circumstances. Local *in vivo* administration of GABA$_B$ antagonists has been reported to reduce the burst-firing of the DA neurons in one study [Tepper et al.,1995] although, interestingly, studies with systemically administered GABA$_B$ antagonists have not provided indications that blocking the GABA$_B$ receptors induces any dramatic effects on DA cell firing in extracellular single unit recording studies in anesthetized rats. [Engberg et al., 1993; Shepard et al., 1995] These results suggest that DA cells are not tonically inhibited via GABA$_B$ receptors; however, it is not clear whether these antagonists are sufficiently potent when administered systemically or whether subsets of GABA$_B$ receptors exist which are less affected by the available antagonists. The cloning of the GABA$_B$ receptor subtype should facilitate the development of more selective and efficacious drugs, and provide further insight into the role of these receptors on the DA neurons and the function of the pathways innervating them.

VI. SUMMARY

A large body of evidence supports the view that GABAergic systems play an important role in regulating the function of midbrain DA neurons. Anatomical evidence indicates that there is a high concentration of GABA in the substantia nigra and a large proportion of the terminals which form synapses with DA neurons contain GABA. In addition, *in vivo* and *in vitro* studies have demonstrated that GABA and GABAergic agents acting through both GABA$_A$ and GABA$_B$ receptors have clear neurophysiological and pharmacological effects on DAergic neuronal activity. These effects appear mediated by both postsynaptic- and presynaptically located GABA receptors. Current research has highlighted the fact that the effects of GABA are complex at many levels. It has been recognized for many years that multiple GABAergic neuronal systems can affect DA function, but more recently, it has also become clear that there is complexity at the cellular level; GABA$_A$ and GABA$_B$ receptors appear to be differentially located post-synaptically with respect to different afferents and, probably, presynaptically, as well. In addition, it is clear that there are multiple functionally important GABA$_A$ receptor subtypes, and likely to be GABA$_B$ subtypes. It is hopeful that future studies will elucidate the specifics of the locations of these receptors and reveal whether different systems (e.g. intrinsic neurons, striatal afferents, pallidal afferents) selectively target different subtypes of the GABA$_A$ and GABA$_B$ receptors, and/or whether these are located on different regions of the DA cells. This information might allow specific manipulation of the different systems in order to understand how they regulate DA neurons. Besides addressing cellular location of receptor subtypes, and relationships between receptor subtypes and the neuronal inputs to DA neurons, it will be important to learn more about presynaptic regulation of neurotransmitter release, and the cellular effects of

GABA on non-GABAergic neurons which regulate DA neurons. As future research begins to sort out the answers to these questions, an important goal is to establish the role of specific GABAergic afferents and specific GABA receptor subtypes in the regulation of the rate and pattern of DA cell firing, the ratio of silent to active dopamine neurons and the response of DA neurons to external stimuli. GABAergic systems have the clear potential for being very important regulators of these processes and could play a critical role in protecting DA cells from the damaging effects of hyperactivity. At issue is the hope that a better understanding of the nature of GABA/DA interactions could enhance potential for therapeutically adjusting critical aspects of DA system function in disease states.

REFERENCES

Aceves, J., Floran, B., Martinez-Fong, D., Benitez, J., Sierra, A. and Flores, G., Activation of D_1 receptors stimulates accumulation of γ-aminobutyric acid in slices of the pars reticulata of 6-hydroxydopamine-lesioned rats, *Neurosci. Lett.*, 145, 40, 1992.

Aghajanian, G. K. and Bunney, B. S., Dopamine "autoreceptors": Pharmacological characterization by microiontophoretic single cell recording studies, *Naunyn-Schmiedeberg's Arch. Pharmacol.*, 297, 1, 1977.

Akaoka, H., Charlety, P., Saunier, C.-F., Buda, M. and Chouvet, G., Inhibition of nigral dopamine neurons by systemic and local apomorphine: possible contribution of dendritic autoreceptors, *Neurosci.*, 49, 879, 1992.

Altar, C.A. and Hauser, K., Topography of substantia nigra innervation by D_1 receptor-containing striatal neurons, *Brain Res.*, 410, 1, 1987.

Anderson, D. R., Li, W. and Tepper, J. M., GABAergic inhibition of nigrostriatal dopaminergic neurons by selective activation of pars reticulata projection neurons, *Soc. Neurosci. Abstr.*, 19, 1432, 1993.

Angelotti, T. P. and Macdonald, R. L., Assembly of $GABA_A$ receptor subunits: $\alpha_1\beta_1$ and $\alpha_1\beta_1\gamma_2$s subunits produce unique ion channels with dissimilar single-channel properties, *J. Neurosci.*, 13, 1429, 1993.

Araki, T., Sato, M., Kiyama, H., Manabe, Y. and Tohyama, M., Localization of $GABA_A$-receptor γ_2-subunit mRNA-containing neurons in the rat central nervous system, *Neurosci.*, 47, 45, 1992.

Arbilla, S., Kamal, L. and Langer, S. Z., Presynaptic GABA autoreceptors on GABAergic nerve endings of the rat substantia nigra, *Europ. J. Pharmacol.*, 57, 211, 1979.

Arnt, J. and Scheel-Krüger, J., GABA in the ventral tegmental area: differential regional effects on locomotion, aggression and food intake after microinjection of GABA agonists and antagonists, *Life Sci.*, 25, 1351, 1979.

Baring, M. D., Walters, J. R. and Eng, N., Action of systemic apomorphine on dopamine cell firing after neostriatal kainic acid lesion, *Brain Res.*, 181, 214, 1980.

Bateson, A. N., Lasham, A. and Darlison, M. G., γ-Aminobutyric acid$_A$ receptor heterogeneity is increased by alternative splicing of a novel β-subunit gene transcript, *J. Neurochem.*, 56, 1437, 1991.

Bedard, P. and Larochelle, L., Effect of section of the strionigral fibers on dopamine turnover in the forebrain of the rat, *Exp. Neurol.*, 41, 314, 1973.

Björklund, A. and Lindvall, O., Dopamine in dendrites of substantia nigra neurons: suggestions for a role in dendritic terminals, *Brain Res.*, 83, 531, 1975.

Bolam, J. P. and Smith, Y., GABA and substance P input to dopaminergic neurones in the substantia nigra of the rat, *Brain Res.*, 529, 57, 1990.

Bonanno, G. and Raiteri, M., Functional evidence for multiple γ-aminobutyric acid$_B$ receptor subtypes in the rat cerebral cortex, *J. Pharmacol. Exp. Ther.*, 262, 114, 1992.

Bonanno, G. and Raiteri, M., γ-aminobutyric acid (GABA) autoreceptors in rat cerebral cortex and spinal cord represent pharmacologically distinct subtypes of the GABA$_B$ receptor, *J. Pharmacol. Exp. Ther.*, 265, 765, 1993a.

Bonanno, G. and Raiteri, M., Multiple GABA$_B$ receptors, *TIPS*, 14, 259, 1993b.

Bowery, N. G., Hudson, A. L. and Price, G. W., GABA$_A$ and GABA$_B$ receptor site distribution in the rat central nervous system, *Neurosci.* 20, 365, 1987.

Bowery, N. G., Knott, C., Moratalla, R. and Pratt, G. D., GABA$_B$ receptors and their heterogeneity, *Adv. Biochem. Psychopharmacol.*, 46, 127, 1990.

Braszko, J. J., Bannon, M. J., Bunney, B. S. and Roth, R. H., Intrastriatal kainic acid: Acute effects on electrophysiological and biochemical measures of nigrostriatal dopaminergic activity, *J. Pharmacol. Exp. Ther.*, 216, 289, 1981.

Brownstein, M. J., Mroz, E. A., Tappaz, M. and Leeman, S. E., On the origin of substance P and glutamic acid decarboxylase (GAD) in the substantia nigra, *Brain Res.*, 135, 315, 1977.

Bunney, B. S. and Aghajanian, G. K., Dopaminergic influence in the basal ganglia: Evidence for striatonigral feedback regulation, in *The Basal Ganglia*, Yahr, M.P. Ed., Raven Press, New York, 1976, 249.

Bunney, B. S. and Aghajanian, G. K., d-Amphetamine-induced depression of central dopamine neurons: evidence for mediation by both autoreceptors and a striato-nigral feedback pathway, *Naunyn-Schmiedeberg's Arch. Pharmacol.*, 304, 255, 1978.

Bunney, B. S. and Grace, A. A., Acute and chronic haloperidol treatment: comparison of effects on nigral dopaminergic cell activity, Life Sciences. 23, 1715, 1978.

Bunney, B. S., Walters, J. R., Roth, R. H. and Aghajanian, G. K., Dopaminergic neurons: Effect of antipsychotic drugs and amphetamine on single cell activity, *J. Pharmacol. Exp. Ther.*, 185, 560, 1973.

Calabresi, P., Mercuri, N. B., De Murtas, M. and Bernardi, G., Involvement of GABA systems in feedback regulation of glutamate- and GABA-mediated synaptic potentials in rat neostriatum, *J. Physiol.*, 440, 581, 1991.

Cameron, D. L. and Williams, J. T., Dopamine D$_1$ receptors facilitate transmitter release, *Nature*, 366, 344, 1993.

Cameron, D.L. and Williams, J.T., Cocaine inhibits GABA release in the VTA through endogenous 5-HT, *J. Neurosci.*, 14, 6763, 1994.

Carlsson, A. and Lindqvist, M., Effect of chlorpromazine and haloperidol on formation of 3-methoxytyramine and normetanephrine in mouse brain, *Acta Pharmacol. Toxicol.*, 20, 140, 1963.

Cash, C. D., Gamma-hydroxybutyrate: an overview of the pros and cons for its being a neurotransmitter and/or a useful therapeutic agent, *Neurosci. Biobehav. Rev.*, 18, 291, 1994.

Collingridge, G. L. and Davies, J., The influence of striatal stimulation and putative neurotransmitters on identified neurones in the rat substantia nigra, *Brain Res.*, 212, 345, 1981.

Crunelli, V. and Leresche, N., A role for GABA$_B$ receptors in excitation and inhibition of thalamocortical cells, *Trends. Neurosci.*, 14, 16, 1991.

Cutting, G. R., Lu, L., O'Hara, B. F., Kasch, L. M., Montrose-Rafizadeh, C., Donovan, D. M., Shimada, S., Antonarakis, S. E., Guggino, W. B., Uhl, G. R. and Kazazian, J.

H. H., Cloning of the γ-aminobutyric acid (GABA)ρ₁ cDNA: a GABA receptor subunit highly expressed in the retina, *Proc. Natl. Acad. Sci.*, 88, 2673, 1991.

Damlama, M., Bolam, J. P. and Tepper, J. M., Axon collaterals of pars reticulata projection neurons synapse on pars compacta neurons, *Soc. Neurosci. Abstr.*, 19, 1432, 1993.

Deniau, J. M., Kitai, S. T., Donoghue, J. P. and Grofova, I., Neuronal interactions in the substantia nigra pars reticulata through axon collaterals of the projection neurons, *Exp.Brain Res.*, 47, 105, 1982.

Diana, M., Mereu, G., Mura, A., Fadda, F., Passino, N. and Gessa, G., Low doses of gamma-hydroxybutyric acid stimulate the firing rate of dopaminergic neurons in unanesthetized rats, *Brain Res.,* 566, 208, 1991.

Diana, M., Pistis, M., Muntoni, A. and Gessa, G., Heterogenous responses of substantia nigra pars reticulata neurons to gamma-hydroxybutyric acid administration, *Eur. J. Pharmacol.*, 230, 363, 1993.

DiChiara, G., Morelli, M., Porceddu, M. L., and Gessa, G. L., Evidence that nigral GABA mediates behavioural responses elicited by striatal dopamine receptor stimulation, *Life Sci.*, 23, 2045, 1978.

DiChiara, G., Porceddu, M. L., Morelli, M., Mulas, M. L. and Gessa, G. L., Evidence for a GABAergic projection from the substantia nigra to the ventromedial thalamus and to the superior colliculus of the rat, *Brain Res.*, 176, 273, 1979.

Domesick, V. B., The anatomical basis for feedback and feed-forward in the striatonigral system, in *Apomorphine and Other Dopaminomimetics, Basic Pharmacology,* Vol. 1, Gessa, G. L. and Corsini, G.U., Eds., Raven Press, New York, 1987, 171.

Dunlap, K. and Fischbach, G. D., Neurotransmitters decrease the calcium conductance activated by depolarization of embryonic chick sensory neurones, *J. Physiol.*, 317, 519, 1981.

Dunlap, K., Holz, G. G. and Rane, S. G., G proteins as regulators of ion channel function, *TINS*, 10, 241, 1987.

Dutar, P. and Nicoll, R. A., A physiological role for GABA_B receptors in the central nervous system, *Nature*, 332, 156, 1988a.

Dutar, P. and Nicoll, R. A., Pre- and postsynaptic GABA_B receptors in the hippocampus have different pharmacological properties, *Neuron*, 1, 585, 1988b.

Engberg, G., Kling-Petersen, T. and Nissbrandt, H., GABA_B receptor activation alters the firing pattern of dopamine neurons in the rat substantia nigra, *Synapse,* 15, 229, 1993.

Engberg, G. and Nissbrandt, H., γ-Hydroxybutyric (GHBA) induces pacemaker activity and inhibition of substantia nigra dopamine neurons by activating GABA_B-receptors, *Naunyn-Schmiedeberg's Arch. Pharmacol.*, 348, 491, 1993.

Fahn, S. and Côté, L. J., Regional distribution of gamma-aminobutyric acid (GABA) in brain of the rhesus monkey, *J. Neurochem.*, 15, 209, 1968.

Fallon, J. H. and Loughlin, S. E., Substantia nigra, in *The Rat Nervous System, Forebrain and Midbrain*, Vol. 1, Paxinos, G. Ed., Academic Press, Sydney, 1985, 353.

Fallon, J. H., Riley, J. N. and Moore, R. Y., Substantia nigra dopamine neurons: separate populations project to neostriatum and allocortex, *Neurosci. Letts.*, 7, 157, 1978.

Floran, B., Aceves, J., Sierra, A. and Martinez-Fong, D., Activation of D1 dopamine receptors stimulates the release of GABA in the basal ganglia of the rat, *Neurosci. Lett.*, 116, 136, 1990.

Fonnum, F., Grofova, I., Rinvik, E., Storm-Mathisen, J. and Walberg, F., Origin and distribution of glutamate decarboxylase in substantia nigra of the cat, *Brain Res.*, 71, 77, 1974.

Francois, C., Percheron, G., Yelnic, J. and Heyner, S., Demonstration of the existence of small local circuit neurons in the Golgi-stained primate substantia nigra, *Brain Res.*, 172, 160, 1979.

Fritschy, J.-M., Benke, D., Mertens, S., Oertel, W. H., Tachi, T. and Mohler, H., Five subtypes of type A γ-aminobutyric acid receptors identified in neurons by double and triple immunofluorescence staining with subunit-specific antibodies, *Proc. Natl. Acad. Sci.*, 89, 6726, 1992.

Gähwiler, B. H. and Brown, D. A., $GABA_B$-receptor-activated K^+ current in voltage-clamped CA3 pyramidal cells in hippocampal cultures, *Proc. Natl. Acad. Sci.*, 82, 1558, 1985.

Gallimberti, L., Ferri, M., Ferrara, S.D., Fadda, F. and Gessa, G.L. Gamma-hydroxybutyric acid in the treatment of alcohol dependence: A double-blind study, *Alcoholism*, 16, 673, 1992.

Gambarana, C., Beattie, C. E., Rodriguez, A. R. and Siegel, R. E., Region-specific expression of messenger RNAs encoding $GABA_A$ receptor subunits in the developing rat brain, *Neurosci.*, 45, 423, 1991.

Gao, B., Fritschy, J. M., Benke, D. and Mohler, H., Neuron-specific expression of $GABA_A$-receptor subtypes: differential association of the α-1 and α-3 subunits with serotonergic and GABAergic neurons, *Neurosci.*, 54, 881, 1993.

Garcia-Munoz, M., Nicolaou, N. M., Tulloch, I. F., Wright, A. K., Arbuthnott, G. W., Feedback loop or output pathway in striatonigral fibres, *Nature*, 265, 363, 1977.

Geffen, L. B., Jessel, T. M., Cuello, A. C. and Iversen, L. L., Dopamine release in substantia nigra, *Nature*, 260, 257, 1976.

Gerfen, C. R., The neostriatal mosaic. I. Compartmental organization of projections from the striatum to the substantia nigra in the rat, *J. Comp. Neurol.*, 236, 454, 1985.

Gerfen, C. R., Herkenham, M. and Thibault, J., The neostriatal mosaic: II. Patch-and matrix-directed mesostriatal dopaminergic and non-dopaminergic systems, *J. Neurosci.*, 7, 3915, 1987.

German, D. C., Dalsass, M. and Kiser, R. S., Electrophysiological examination of the ventral tegmental (A10) area in the rat, *Brain Res.*, 181, 191, 1980.

Gessa, G. L., Vargiu, L., Crabai, F., Boero, C. C., Caboni, F. and Camba, R., Selective increase of brain dopamine induced by gamma-hydroxybutyrate, *Life Sci.*, 5, 1921, 1966.

Giralt, M. T., Bonnano, G. and Raiteri, M., GABA terminal autoreceptors in the pars compacta and in the pars reticulata of the rat substantia nigra are $GABA_B$, *Eur. J. Pharmacol.*, 175, 137, 1990.

Grace, A. A. and Bunney, B. S., Paradoxical GABA excitation of nigral dopaminergic cells: Indirect mediation through reticulata inhibitory interneurons, *Eur. J. Pharmacol.*, 59, 211, 1979.

Grace, A. A. and Bunney, B. S., Effects of baclofen on nigral dopaminergic cell activity following acute and chronic haloperidol treatment, *Brain Res. Bull.*, 5, 537, 1980.

Grace, A. A. and Bunney, B. S., Intracellular and extracellular electrophysiology of nigral dopaminergic neurons - 1. Identification and characterization, *Neurosci.*, 10, 301, 1983a.

Grace, A. A. and Bunney, B. S., Intracellular and extracellular electrophysiology of nigral dopaminergic neurons - II. Action potential generating mechanisms and morphological correlates, *Neurosci.*, 10, 317, 1983b.

Grace, A. A. and Bunney, B. S., The control of firing pattern in nigral dopamine neurons: Single spike firing, *J. Neurosci.*, 4, 2866, 1984.

Grace, A. A. and Bunney, B. S., Opposing effects of striatonigral feedback pathways on midbrain dopamine cell activity, *Brain Res.*, 333, 271, 1985.

Grace, A. A. and Onn, S.-P., Morphology and electrophysiological properties of immunocytochemically identified rat dopamine neurons recorded in vitro, *J. Neurosci.*, 9, 3463, 1989.

Grofová, I., The identification of striatal and pallidal neurons projecting to substantia nigra. An experimental study by means of retrograde axonal transport of horseradish peroxidase, *Brain Res.*, 91, 286, 1975.

Grofová, I., Deniau, J. M. and Kitai, S. T., Morphology of the substantia nigra pars reticulata projection neurons intracellularly labeled with HRP, *J. Comp. Neurol.*, 208, 352, 1982.

Grofová, I. and Rinvik, E., An experimental electron microscopic study on the striatonigral projection in the cat, *Exp. Brain Res.*, 11, 249, 1970.

Groves, P. M., Wilson, C. J., Young, S. J. and Rebec, G. V., Self-inhibition by dopaminergic neurons, *Science*, 190, 522, 1975.

Gulley, R. L. and Wood, R. L., The fine structure of the neurons in the rat substantia nigra, *Tiss. Cell.*, 3, 675, 1972.

Guyenet, P. G., Agid, Y., Javoy, F., Beaujouan, J. C., Rossier, J. and Glowinski, J., Effects of dopaminergic receptor agonists and antagonists on the activity of the neostriatal cholinergic system, *Brain Res.*,84, 227, 1975.

Hajós, M. and Greenfield, S. A., Topographic heterogeneity of substantia nigra neurons: diversity in intrinsic membrane properties and synaptic inputs, *Neurosci.*, 55, 919, 1993.

Hattori, T., Fibiger, H. C. and McGeer, P. L., Demonstration of a pallido-nigral projection innervating dopaminergic neurons, *J. Comp. Neurol.*, 162, 487, 1975.

Hattori, T., McGeer, P. L., Fibiger, H. C. and McGeer, E. G., On the source of GABA-containing terminals in the substantia nigra. Electron microscopic autoradiographic and biochemical studies, *Brain Res.*, 54, 103, 1973.

Häusser, M.A. and Yung, W.H., Inhibitory synaptic potentials in guinea-pig substantia nigra dopamine neurons *in vitro*, *J. Physiol.*, 479, 401, 1994.

Herb, A., Wisden, W., Lüddens, H., Puia, G., Vicini, S. and Seeburg, P. H., The third γ subunit of the γ-aminobutyric acid type A receptor family, *Proc. Natl. Acad. Sci.*, 89, 1433, 1992.

Holz, G. G., IV, Rane, S. G. and Dunlap, K., GTP-binding proteins mediate transmitter inhibition of voltage-dependent calcium channels, *Nature*, 319, 670, 1986.

Hommer, D. W. and Bunney, B. S., Effect of sensory stimuli on the activity of dopaminergic neurons: Involvement of non-dopaminergic nigral neurons and striato-nigral pathways, *Life Sci.*, 27, 377, 1980.

Iwatsubo, K. and Clouet, D. H., Effects of morphine and haloperidol on the electrical activity of rat nigrostriatal neurons, *J. Pharmacol. Exp. Ther.*, 202, 429, 1977.

Jessell, T. M., Emson, P. C., Paxinos, G. and Cuello, A. C., Topographic projections of substance P and GABA pathways in the striato- and pallido-nigral system: a biochemical and immunohistochemical study, *Brain Res.*, 152, 487, 1978.

Jiang, Z.-G., Pessia, M. and North, R. A., Dopamine and baclofen inhibit the hyperpolarization-activated cation current in rat ventral tegmental neurones, *J. Physiol.*, 462, 753, 1993.

Johnson, S. W., Mercuri, N. B. and North, R. A., 5-hydroxytryptamine$_{1B}$ receptors block the GABA$_B$ synaptic potential in Rat dopamine neurons, *J. Neurosci.*, 12, 2000, 1992.

Johnson, S. W. and North, R. A., Opioids excite dopamine neurons by hyperpolarization of local interneurons, *J. Neurosci.*, 12, 483, 1992a.

Johnson, S. W. and North, R. A., Two types of neurons in the rat ventral tegmental area and their synaptic inputs, *J. Physiol.*, 450, 455, 1992b.

Johnston, G. A. R., GABA receptors: As complex as ABC?, *Clin. Exp. Pharmacol. Physiol.,* 21, 521, 1994.

Kataoka, K., Bak., I. J., Hassler, R., Kim, J. S. and Wagner, A., L-Glutamate decarboxylase and choline acetyltransferase activity in the substantia nigra and the striatum after surgical interruption of the strio-nigral fibres of the baboon, *Exp. Brain Res.,* 19, 217, 1974.

Kayadjanian, N., Gioanni, H., Menetrey, A. and Besson, M. J., Muscarinic receptor stimulation increases the spontaneous [3H]GABA release in the rat substantia nigra through muscarinic receptors localized on striatonigral terminals, *Neurosci.,* 63, 989, 1994.

Kehr, W., Carlsson, A., Lindqvist, M., Magnusson, T. and Atack, C., Evidence for a receptor-mediated feedback control of striatal tyrosine hydorxylase, *J. Pharm. Pharmacol.,* 24, 744, 1972.

Khrestchatisky, M., MacLennan, A. J., Chiang, M.-Y., Xu, W., Jackson, M. B., Brecha, N., Sternini, C., Olsen, R. W. and Tobin, A. J., A novel α subunit in rat brain GABA$_A$ receptors, *Neuron,* 3, 745, 1989.

Kilpatrick, I. C., Starr, M. S., Fletcher, A., James, T. A. and MacLeod, N. K., Evidence for a GABAergic nigrothalamic pathway in the rat. I. Behavioral and biochemical studies, *Exp. Brain Res.,* 40, 45, 1980.

Kim, J. S., Bak, I. J., Hassler, R. and Okada, Y., Role of gamma-aminobutyric acid (GABA) in the extrapyramidal motor system. 2. Some evidence for the existence of a type of GABA-rich strio-nigral neurons, *Exp. Brain Res.,* 14, 95, 1971.

Knoflach, F., Rhyner, T., Villa, M., Kellenberger, S., Drescher, U., Malherbe, P., Sigel, E. and Mohler, H., The gamma 3-subunit of the GABA$_A$ receptor confers sensitivity to benzodiazepine receptor ligands, *FEBS Letts.,* 293, 191, 1991.

Knott, C., Maguire, J. J., Moratalla, R. and Bowery, N. G., Regional effects of pertussis toxin *in vivo* and *in vitro* on GABA$_B$ receptor binding in rat brain, *Neurosci.,* 52, 73, 1993.

Korf, J., Zieleman, M., and Westerink, B. H. C., Dopamine release in substantia nigra?, *Nature,* 260, 257, 1976.

Laborit, H., Buchard, F., Larobit, G., Kind, A. and Weber, B., Emploie du 4-hydroxybutyrate de Na en anesthesic et en reamination, *Agressologie,* 1, 549, 1960.

Lacey, M. G., Mercuri, N. B. and North, R. A., Dopamine acts on D$_2$ receptors to increase potassium conductance in neurones of the rat substantia nigra zona compacta, *J. Physiol.,* 392, 397, 1987.

Lacey, M. G., Mercuri, N. B. and North, R. A., On the potassium conductance increase activated by GABA$_B$ and dopamine D$_2$ receptors in rat substantia nigra neurones, *J. Physiol.,* 401, 437, 1988.

Lacey, M. G., Mercuri, N. B. and North, R. A., Two cell types in rat substantia nigra zona compacta distinguished by membrane properties and the actions of dopamine and opioids, *J. Neurosci.,* 9, 1233, 1989.

Ladinsky, H., Consolo, S., Bianchi, S. and Jori, A., Increase in striatal acetylcholine by picrotoxin in the rat: evidence for a gabaergic-dopaminergic-cholinergic link, *Brain Research,* 108, 351, 1976.

LaHoste, G.J. and Marshall, J.F., Nigral D$_1$ and striatal D$_2$ receptors mediate the behavioral effects of dopamine agonists, *Behav. Brain Res.,* 38, 233, 1990.

Lasham, A., Vreugdenhil, E., Bateson, A. N., Barnard, E. A. and Darlison, M. G., Conserved organization of γ-aminobutyric acid$_A$ receptor genes: cloning and analysis of the chicken β4-subunit gene, *J. Neurochem.,* 57, 352, 1991.

Levitan, E. S., Blair, L. A. C., Dionne, V. E. and Barnard, E. A., Biophysical and pharmacological properties of cloned GABA$_A$ receptor subunits expressed in xenopus oocytes, *Neuron,* 1, 773, 1988a.

Levitan, E. S., Schofield, P. R., Burt, D. R., Rhee, L. M., Wisden, W., Köhler, M., Fujita, N., Rodriguez, H. F., Stephenson, A., Darlison, M. G., Barnard, E. A. and Seeburg, P. H., Structural and functional basis for GABA$_A$ receptor heterogeneity, *Nature,* 335, 76, 1988b.

Llinas,R., Greenfield, S. A. and Jahnsen, H., Electrophysiology of pars compacta cells in the in vitro substantia nigra - a possible mechanism for dendritic release, *Brain Res.,* 194, 127, 1984.

Lo, M. M., Niehoff, D. L., Kuhar, M. J. and Snyder, S. H., Differential localization of type I and type II benzodiazepine binding sites in substantia nigra, *Nature,* 306, 57, 1983.

Lolait, S. J., O'Carroll, A.-M., Kusano, K., Muller, J.-M., Brownstein, M. J. and Mahan, L. C., Cloning and expression of a novel rat GABA$_A$ receptor, *FEBS Letts.,* 246, 145, 1989.

Lüddens, H., Pritchett, D. B., Köhler, M., Killisch, I., Keinänen, K., Monyer, H., Sprengel, R. and Seeburgh, P. H., Cerebellar GABA$_A$ receptor selective for a behavioral alcohol antagonist, *Nature,* 346, 648, 1990.

Lüddens, H. and Wisden, W., Function and pharmacology of multiple GABA$_A$ receptor subunits, TIPS, 12, 49, 1991.

MacLennan, A. J., Brecha, N., Khrestchatisky, M., Sternini, C., Tillakaratne, N. J. K., Chiang, M.-Y., Anderson, K., Lai, M. and Tobin, A. J., Independent cellular and ontogenetic expression of mRNAs encoding threeα polypeptides of the rat GABA$_A$ receptor, *Neurosci.,* 43, 369, 1991.

MacLeod, N. K., James, T. A., Kilpatrick, I. C. and Starr, M. S., Evidence for a GABAergic nigrothalamic pathway in the rat. II. Electrophysiological studies, *Exp. Brain Res.,* 40, 55, 1980.

MacNeil, D., Gower, M. and Szymanska, I., Response of dopamine neurons in substantia nigra to muscimol, *Brain Res.,* 154, 401, 1978.

Malherbe, P., Sigel, E., Baur, R., Persohn, E., Richards, J. G. and Möhler, H., Functional expression and sites of gene transcription of a novel α subunit of the GABA$_A$receptor in rat brain, *FEBS Letts.,* 260, 261, 1990.

Marshall, J.R. and Ungerstedt, U., Striatal efferent fibers play a role in maintaining rotational behavior in the rat. *Science,* 198, 62, 1977.

McMillen, B. A., Williams, H. L., Lehmann, H. and Shepard, P.D., On central muscle relaxants, strychnine-insensitive glycine receptors and two old drugs: zoxazolamine and HA-966, *J. Neural Transm.,* 89, 11, 1992.

Mendez, I., Elisevich, K. and Flumerfelt, B. A., GABAergic synaptic interactions in the substantia nigra, *Brain Res.,* 617, 1993.

Mereu, G. and Biggio, G., Effects of agonists, inverse agonists, and antagonists of benzodiazepine receptors on the firing rates of substantia nigra pars reticulata neurons, *Adv. Biochem.,* 38, 201, 1983.

Mereu, G., Carcangiu, G., Concas, A., Passino, N. and Biggio, G., Reduction of reticulata neuronal activity by zolpidem and alpidem, two imidazopyridines with high affinity for type I benzodiazepine receptors, *Eur. J. Pharmacol.,* 179, 339, 1990.

Mereu, G., Costa, E., Livsey, C. T. and Vicini, S., Modulation of GABA$_A$-gated Cl⁻ currents in nigral neurons in slices, in *GABAergic Synaptic Transmission, Advances in Biochemical Psycholpharmacology,* Vol. 47, Biggio, G., Concas, A. and Costa, E., Eds., Raven Press, New York, 1992, 179.

Mereu, G., Fadda, F. and Gessa, G.L., Ethanol stimulates the firing rate of nigral dopaminergic neurons in unanesthetized rats, *Brain Res.,* 292, 63, 1984.

Mereu, G., Muntoni, F., Calabresi, P., Romani, F., Boi, V. and Gessa, G.L., Responsiveness to 'autoreceptor' doses of apomorphine is inversely correlated with the firing rate of dopaminergic A9 neurons: action of baclofen, *Neurosci. Lett.*, 65, 161, 1986.

Morrisett, R. A., Mott, D.D., Lewis, D.V., Swartzwelder, H.S. and Wilson, W.A., GABA$_A$ -receptor-mediated inhibition of the N-methyl-D-aspartate component of synaptic transmission in the rat hippocampus, *J. Neurosci.*, 11, 203, 1991.

Mott, D.D. and Lewis, D.V., The pharmacology and function of central GABA$_B$ receptors, *Int. Rev. Neurobiol.*, 36, 97, 1994.

Mueller, A. L. and Brodie, M. S., Intracellular recording from putative dopamine-containing neurons in the ventral tegmental area of Tsai in a brain slice preparation, *J. Neurosci. Meth.*, 28, 15, 1989.

Nakamura, S., Tsai, C. T. and Iwama, K., Antagonizing effects of haloperidol and bicuculline on inhibition of neurons of the substantia nigra, pars compacta, *Exp. Neurol.*, 66, 682, 1981.

Nayeen, N., Green, T.P., Martin, I. L. and Bernard, E. A., Quaternary structure of the native GABA$_A$ receptor determined by electron microscopic image analysis, *J. Neurochem.*, 62, 815, 1994.

Newberry, N. R. and Nicoll, R. A., Direct hyperpolarizing action of baclofen on hippocampal pyramidal cells, *Nature*, 308, 450, 1984.

Nicholson, L. F. B., Faull, R. L. M., Waldvogel, H. J. and Dragunow, M., The regional, cellular and subcellular localization of GABA$_A$/benzodiazepine receptors in the substantia nigra of the rat, *Neurosci.*, 50, 355, 1992.

Nieoullon, A., Cheramy, A., Leviel, V. and Glowinsky, J., Effects of the unilateral nigral application of dopaminergic drugs on the in vivo release of dopamine in the two caudate nuclei of the cat, *Eur. J. Pharmacol.*, 53, 289, 1977.

Nitsch, C. and Riesenberg, R., Immunocytochemical demonstration of GABAergic synaptic connections in rat substantia nigra after different lesions of the striatonigral projection, *Brain Res.*, 461, 127, 1988.

Okada, Y., Nitsch-Hassler, C., Kim, J. S., Bak, I. J. and Hassler, R., Role of γ-aminobutyric acid (GABA) in the extrapyramidal motor system. I Regional distribution of GABA in rabbit, rat, guinea pig and baboon CNS, *Exp. Brain Res.*, 13, 514, 1971.

Olianas, M.C., De Montis, G., Mulas, G. and Tagliamonte, A., The striatal dopaminergic function is mediated by the inhibition of a nigra, non-dopaminergic neuronal system via a strio-nigral GABAergic pathway, *Eur. J. Pharmacol.*, 49, 233, 1978.

Olpe, H.-R., Koelle, W. P., Wolf, P. and Haas, H. L., The action of baclofen on neurons of the substantia nigra and of the ventral tegmental area, *Brain Res.*, 134, 577, 1977.

Paden, C., Wilson, C. J., and Groves, P. M., Amphetamine-induced release of dopamine from the substantia nigra in vitro, *Life Sci.*, 19, 1499, 1976.

Pan, H. S., Penney, J. B., Jr. and Young, A. B., Characterization of benzodiazepine receptor changes in substantia nigra, globus pallidus and entopeduncular nucleus after striatal lesions, *J. Pharmacol. Exp. Ther.*, 230, 768, 1984.

Pérez de la Mora, M., Fuxe, K., Hökfelt, T. and Ljungdahl, Å., Effect of apomorphine on the GABA turnover in the DA cell group rich area of the mesencephalon. Evidence for the involvement of an inhibitory gabargic feedback control of the ascending DA neurons, *Neurosci. Lett.*, 1, 109, 1975.

Pérez de la Mora, M., Fuxe, K., Hökfelt, T. and Ljungdahl, Å., Further evidence that apomorphine increases GABA turnover in the DA cell body rich and DA nerve terminal rich areas of the brain, *Neurosci. Lett.*, 2, 239, 1976.

Pérez de la Mora, M. and Fuxe, K., Brain GABA, dopamine and acetylcholine interactions. 1. Studies with oxotremorine, *Brain Res.*, 135, 107, 1977.

183

Persohn, E., Malherbe, P. and Richards, J. G., Comparative molecular neuroanatomy of cloned GABA$_A$ receptor subunits in the rat CNS, *J. Comp. Neurol.*, 326, 193, 1992.

Pessia, M., Jiang, Z.-G., North, R. A. and Johnson, S. W., Actions of 5-hydroxytryptamine on ventral tegmental area neurons of the rat in vitro, *Brain Res.*, 654, 324, 1994.

Pinnock, R.D., Sensitivity of compacta neurones in the rat substantia nigra to dopamine agonists, *Eur. J. Pharmacol.*, 96, 269, 1983.

Pinnock, R. D., Hyperpolarizing action of baclofen on neurons in the rat substantia nigra slice, *Brain Res.*, 322, 337, 1984.

Porceddu, M.L., Giorgi, O., Ongini, E., Mele, S. and Biggio, G., ^{3}H-SCH 23390 binding sites in the rat substantia nigra: Evidence for a presynaptic localization and innervation by dopamine, *Life Sci.*, 39, 321, 1986.

Precht, W. and Yoshida, M., Blockage of caudate-evoked inhibition of neurons in the substantia nigra by picrotoxin, *Brain Res.*, 32, 229, 1971.

Preston, R. J., McCrea, R. A., Chang, H. T. and Kitai, S. T., Anatomy and physiology of substantia nigra and retrorubral neurons studies by extra- and intracellular recording and by horseradish peroxidase labeling, *Neurosci.*, 6, 331, 1981.

Pritchett, D. B., Lüddens, H. and Seeburg, P. H., Type I and Type II GABA$_A$-benzodiazepine receptors produced in transfected cells, *Science*, 245, 1389, 1989a.

Pritchett, D. B. and Seeburg, P. H., γ-aminobutyric acid$_A$ receptor α5-subunit creates novel type II benzodiazepine receptor pharmacology, *J. Neurochem.*, 54, 1802, 1990.

Pritchett, D. B., Sontheimer, H., Shivers, B. D., Ymer, S., Kettenmann, H., Schofield, P. R. and Seeburg, P. H., Importance of a novel GABA$_A$ receptor subunit for benzodiazepine pharmacology, *Nature*, 338, 582, 1989b.

Reiner, A. and Anderson, K. D., The patterns of neurotransmitter and neuropeptide co-occurrence among striatal projection neurons: Conclusion based on recent findings, *Brain Res. Bull.*, 15, 251, 1990.

Reisine, T. D., Nagy, J. E., Beaumont, K., Fibiger, H. C. and Yamamura, H. I., The localization of receptor binding sites in the substantia nigra and striatum of the rat, *Brain Res.*, 177, 241, 1979.

Reubi, J.C., Iversen, L.K., and Jessell, T.M., Dopamine selectively increases ^{3}H-GABA release from slices of rat substantia nigra *in vitro*, *Nature*, 268, 652, 1977.

Ribak, C. E., Vaughn, J. E., Saito, K., Barber, R. and Roberts, E., Immunocytochemical localization of glutamate decarboxylase in rat substantia nigra, *Brain Res.*, 116, 287, 1976.

Ribak, C. E., Vaughn, J. E., and Roberts, E., GABAergic nerve terminals decrease in substantia nigra following hemitransection of striatonigral and pallido-nigral pathways, *Brain Res.*, 192, 413, 1980.

Rick, C. E., and Lacey, M. G., Rat substantia nigra pars reticulata neurons are tonically inhibited via GABA$_A$, but not GABA$_B$, receptors in vitro, *Brain Res.*, 659, 133, 1994.

Robertson, G.S. and Robertson, H.A., D$_1$ and D$_2$ dopamine agonist synergism: Separate sites of action? *TIPS*, 8, 295, 1987.

Roeper, J., Hainsworth, A. H. and Ashcroft, F. M., Tolbutamide reverses membrane hyperpolarization induced by activation of D$_2$ receptors and GABA$_B$ receptors in isolated substantia nigra neurones, *Pflugers Archiv. - Eur. J. Physiol.*, 416, 473, 1990.

Ross, R. J., Waszczak, B. L., Lee, E. K. and Walters, J. R., Effects of benzodiazepines on single unit activity in the substantia nigra pars reticulata, *Life Sci.*, 31, 1025, 1982.

Roth, R. H. and Giarman, N. J., γ-Butyrolactone and γ-hydroxtbutyric acid - I. Distribution and metabolism, *Biochem. Pharmacol.*, 15, 13333, 1966.

Roth, R. H., Walters, J. R. and Aghajanian, G. K., Effect of impulse flow on the release and synthesis of DA in the rat striatum, in *Frontiers in Catecholamine Research*, Snyder, S.H. and Consta, E., Eds., Pergamon Press, New York, 1973, 567.

Rubin, B A. and Giarman, N. J., The therapy of experimental influenza in mice with antibiotic lactones and related compounds, *Yale J. Biol. Med.*, 19, 1017, 1947.

Scarnati, E., Forchetti, C., Ciancarelli, G., Pacitti, C. and Agnoli, A., Responsiveness of nigral neurons to the stimulation of striatal dopaminergic receptors in the rat, *Life Sci.*, 26, 1203, 1980.

Schofield, P. R., Darlison, M. G., Fujita, N., Burt, D. R., Stephenson, F. A., Rodriguez, H., Rhee, L. M., Ramachandran, J., Reale, V., Glencorse, T. A., Seeburg, P. H. and Barnard, E. A., Sequence and functional expression of the GABA$_A$ receptor shows a ligand-gated receptor super-family, *Nature,* 328, 221, 1987.

Schwynn, R. C. and Fox, C. A., The primate substantia nigra: a Golgi and electron microscopic study, *J. Hirnforsch.*, 16, 95, 1972.

Seutin, V., Johnson, S. W. and North, R. A., Effect of dopamine and baclofen on *N*-methyl-D-aspartate-induced burst firing in rat ventral tegmental neurons, *Neurosci.*, 58: 201, 1994.

Shepard, P. D., Connelly, S. T., Lehrmann, H. and Grobaski, K. C., Effects of the enantiomers of (±)-HA-966 on dopamine neurons: an electrophysiological study of a chiral molecule, *Europ. J. Pharmacol.,* in press, 1995.

Shepard, P. D. And Lehmann, H., (±)-1-Hydroxy-3-aminopyrrilidone-2 (HA-966) inhibits the activity of substantia nigra dopamine neurons through a non-N-methyl-D-aspartate receptor-mediated mechanism, *J. Pharmacol. Exp. Ther.*, 261, 387, 1992.

Shivers, B. D., Killisch, I., Sprengel, R., Sontheimer, H., Köhler, M., Schofield, P. R. and Seeburg, P. H., Two novel GABA$_A$ receptor subunits exist in distinct neuronal subpopulations, *Neuron*, 3, 327, 1989.

Singh, L., Donald, A.E., Foster, A.C., Hutson, P.H., Ivensen, L.L., Iversen, S.D., Kemp, J.A., Leeson, P.D., Marshall, G.R., Oles, R.J., Priestley, T., Thorn, L., Tricklebank, M.D., Vass, C.A. and Williams, B.J., Enantiomers of HA-966 (3-amino-1-hydroxypyrrolid-2-one) exhibit distinct central nervous system effects: (+)-HA-966 is a selective glycine/N-methyl-D-aspartate receptor antagonist, but (-)-HA-966 is a potent gamma-butyrolactone-like sedative., *Proc. Natl. Acad. Sci.*, 87, 347, 1990.

Sladek, J.R. and Parnavelas, J.G., Catecholamine-containing dendrites in primate brain, *Brain Res.*, 100, 657, 1975.

Smith, Y. and Bolam, J. P., Neurons of the substantia nigra reticulata receive a dense GABA-containing input from the globus pallidus in the rat, *Brain Res.*, 493, 160, 1989.

Smith, Y. and Bolam, J. P., The output neurones and the dopaminergic neurones of the substantia nigra receive a GABA-containing input from the globus pallidus in the rat, *J. Comp. Neurol.*, 296, 47, 1990.

Smith, Y. and Bolam, J. P., Convergence of synaptic inputs from the striatum and the globus pallidus onto identified nigrocollicular cells in the rat: a double anterograde labelling study, *Neurosci.*, 44, 45, 1991.

Somogyi, P., Bolam, J. P. and Smith, A. D., Monosynaptic cortical input and local axon collaterals of identified striatonigral neurons. A light and electron microscopic study using the golgi-peroxidase transport-degeneration procedure, *J. Comp. Neurol.*, 195, 567, 1981.

Somogyi, P., Hodgson, A. J. and Smith, A. D., An approach to tracing neuron networks in the cerebral cortex and basal ganglia. Combination of Golgi staining, retrograde transport of horseradish peroxidase and anterograde degeneration of synaptic boutons in the same material, *Neurosci.*, 4, 1805, 1979.

Staley, K. J., Soldo, B. L. and Proctor, W. R., Ionic mechanisms of neuronal excitation by inhibitory GABA$_A$ receptors, *Science*, 269, 977, 1995.

Sugita, S., Johnson, S. W. and North, R. A., Synaptic inputs to GABA$_A$ and GABA$_B$ receptors originate from discrete afferent neurons, *Neurosci. Letts.*, 134, 207, 1992.

Tappaz, M. L., Brownstein, M. J. and Palkovits, M., Distribution of glutamate decarboxylase in discrete brain nuclei, *Brain Res.*, 108, 371, 1976.

Tepper, J. M., Sawyer, S. F. and Groves, P. M., Electrophysiologically identified nigral dopaminergic neurons intracellularly labeled with HRP: Light-microscopic analysis, *J. Neurosci.*, 7, 2794, 1987.

Tepper, J. M., Martin, L. P. and Anderson, D. R., GABA$_A$ receptor-mediated inhibition of rat substantia nigra dopaminergic neurons by pars reticulata projection neurons, *J. Neurosci.*, 15, 3092, 1995.

Timmerman, W., Zwaveling, J. and Westerink, B.H.C., Dopaminergic modulation of the GABA release in the substantia nigra reticulata. In *Monitoring Molecules in Neuroscience*, Rollema, H., Westerink, B. and Drijfhout, W.J., Eds., Groningen, Univ. Of Groningen, 105, 1991.

Trabucchi, M., Cheney, D.L., Racagni, G. and Costa, E., In vivo inhibition of striatal acetylcholine turnover by L-DOPA, morphine and (+)-amphetamine, *Brain Res.*, 85, 130, 1975.

Ueki, A., Uno, M., Anderson, M. and Yoshida, M., Monosynaptic inhibition of thalamic neurons produced by stimulation of the substantia nigra, *Experientia*, 33, 1480, 1977.

Van Den Pol, A. N., Smith, A. D. and Powell, J. F., GABA axons in synaptic contact with dopamine neurons in the substantia nigra: double immunocytochemistry with biotin-peroxidase and protein A-colloidal gold, *Brain Res.*, 348, 146, 1985.

Vincent, S. R., Hattori, T. and McGeer, E. G., The nigrotectal projection: a biochemical and ultrastructural characterization, *Brain Res.*, 151, 159, 1978.

Von Krosigk, M., Smith, Y., Bolam, J. P. and Smith, A. D., Synaptic organization of GABAergic inputs from the striatum and the globus pallidus onto neurons in the substantia nigra and retrorubral field which project to the medullary reticular formation, *Neurosci.*, 50, 531, 1992.

Waddington, J. L. and Cross, A. J., Neurochemical changes following kainic acid lesions of the nucleus accumbens: Implications for a GABAergic accumbal-ventral tegmental pathway, *Life Sci.*, 22, 1011, 1978.

Walaas, I. and Fonnum, F., Biochemical evidence for γ-aminobutyrate containing fibers from the nucleus accumbens to the substantia nigra and ventral tegmental area in the rat, *Neurosci.*, 5, 63, 1980.

Waldmeier, P. C., The GABAB antagonist, CGP 35348, antagonizes the effects of baclofen, γ-butyrolactone and HA-966 on rat striatal dopamine synthesis, *Naunyn-Schmiedeberg's Arch. Pharmacol.*, 343, 173, 1991.

Walters, J.R., Aghajanian, G. K. Aand Roth, R.H., Dopaminergic neurons: Inhibition of firing by γ-hydroxybutyrate, *Proceedings of the Fifth Internatinoal Congress of Pharmacology*, p. 246, 1972.

Walters, J. R. and Lakoski, J. M., Effect of muscimol on single unit activity of substantia nigra dopamine neurons, *Europ. J. Pharmacol.*, 47, 469, 1978.

Walters, J. R. and Roth, R. H., Dopaminergic neurons: drug induced antagonism of the increase in tyrosine hydorylase activity produced by cessation of impulse flow, *J. Pharmac. Exp. Ther.*, 191, 82, 1974.

Walters, J. R. and Roth, R. H., Dopaminergic neurons: An in vivo system for measuring drug interactions with presynaptic receptors, *Naunyn-Schmiedeberg's Arch. Pharmacol.*, 274, 5, 1976.

Walters, J. R. and Roth, R. H., γ-Hydroxybutrate: considerations of endogenous role and therapeutic potential in *Neurotransmitters and Hypotehses of Neurological Disorders,* Usdin, E., Barcus, J., and Hamburg, D., Eds, Oxford Press, N.Y., 1977, 403.

Walters, J.R., Roth, R. H., and Aghajanian, G. K., Dopaminergic neurons: Similar biochemical and histochemical effects of gamma-hydroxybutyrate and acute lesions of the nigroneostriatal pathway, *J. Pharmacol. Exp. Ther.,* 186, 630, 1973.

Wang, R.Y., Dopaminergic neurons of the rat ventral tegmental area. II. Evidence for autoregulation, *Brain Res. Rev.,* 3, 141, 1981a.

Wang, R.Y., Dopaminergic neurons of the rat ventral tegmental area. III. Effects of d- and l-amphetamine, *Brain Res. Rev.,* 3, 153, 1981b.

Wasseff, M., Berod, A. and Sotelo, C., Dopaminergic dendrites in the pars reticulata of the rat substantia nigra and their striatal input. Combined immunocytochemical localization of tyrosine hydroxylase and anterograde degeneration, *Neurosci.,* 6, 2125, 1981.

Waszczak, B. L., Diazepam potentiates GABA, but not adenosine-mediated inhibition of neurons of the nigral pars reticulata, *Neuropharamcol.,* 22, 953, 1983.

Waszczak, B. L., Bergstrom, D. A. and Walters, J. R., Single unit responses of substantia nigra and globus pallidus neurons to GABA agonist and antagonist drugs, in *GABA and the Basal Ganglia,* Di Chiara, G. And Gessa, G. L., Eds., Raven Press, New York, 1981, 79.

Waszczak, B. L., Eng, N. and Walters, J. R., Effects of muscimol and picrotoxin on single unit activity of substantia nigra neurons, *Brain Res.,* 188, 185, 1980.

Waszczak, B. L., Hume, C. and Walters, J. R., Supersensitivity of substantia nigra pars reticulata neurons to GABAergic drugs after striatal lesions, *Life Sci.,* 28, 2411, 1981.

Waszczak, B.L., Lee, E.K. and Walters, J.R., Effects of anticonvulsant drugs on substantia nigra pars reticulata neurons, *J. Pharmacol. Exp. Ther.,* 239, 606, 1986.

Waszczak, B. L. and Walters, J. R., Effects of GABA-mimetics on substantia nigra neurons, in *Advances in Neurology,* Vol. 23, Chase, T.N. et al, Eds, Raven Press, New York, 1979, pp. 727-739.

Waszczak, B. L. and Walters, J. R., Effects of GABAergic drugs on single unit activity of A9 and A10 dopamine neurons, *Brain Res. Bull.,* 5, Suppl. 2, 465, 1980a.

Waszczak, B. L. and Walters, J. R., Intravenous GABA agonist administration stimulates firing of A10 dopaminergic neurons, *Eur. J. Pharmacol.,* 66, 141, 1980b.

White, F. J. and Wang, R.Y., Comparison of the effects of chronic haloperidol treatment on A9 and A10 dopamine neurons in the rat., *Life Sci.,* 32, 983, 1983.

Whiting, P., McKernan, R. M. and Iversen, L. L., Another mechanism for creating diversity in γ-aminobutyrate type A receptors: RNA splicing directs expression of two forms of γ2 subunit, one of which contains a protein kinase C phosphorylation site, *Proc. Natl. Acad. Sci.,* 87, 9966, 1990.

Wilson-Shaw, D., Robinson, M., Gambarana, C., Siegel, R. E. and Sikela, J. M., A novel γ subunit of the $GABA_A$ receptor identified using the polymerase chain reaction, *FEBS Letts.,* 284, 211, 1991.

Wisden, W., Herb, A., Wieland, H., Keinänen, K., Lüddens, H. and Seeburgh, P. H., Cloning, pharmacological characteristics and expression pattern of the rat $GABA_A$ receptor α4 subunit, *FEBS Letts.,* 289, 227, 1991.

Wisden, W., Laurie, D. J., Monyer, H. and Seeburgh, P. H., The distribution of 13 $GABA_A$ receptor subunit mRNAs in the rat brain. I. Telencephalon, diencephalon, mesencephalon, *J. Neurosci.,* 12, 1040, 1992.

Wu, Y.-N., Mercuri, N. A. and Johnson, S. W., Presynaptic inhibition of γ-Aminobutyric $Acid_B$-mediated synaptic current by adenosine recorded *in vitro* in midbrain dopamine neurons, *J. Pharmacol. Exp. Ther.,* 273, 576, 1995.

Xie, X. and Smart, T.G., Gamma-hydroxybutyrate depresses monosynaptic excitatory and inhibitory postsynaptic potentials in rat hippocampal slices, *Eur. J. Pharmacol.,* 223, 193, 1992.

Yim, C.Y. and Mogenson, G. J., Electrophysiological studies of neurons in the ventral tegmental area of Tsai, *Brain Res.,* 181, 301, 1980.

Ymer, S., Draguhn, A., Köhler, M., Schofield, P. R. and Seeburg, P. H., Sequence and expression of a novel $GABA_A$ receptor α subunit, *FEBS Letts.,* 258, 119, 1989a.

Ymer, S., Schofield, P. R., Draguhn, A., Werner, P., Köhler, M. and Seeburg, P. H., $GABA_A$ receptor subunit heterogeneity: functional expression of cloned cDNAs, *EMBO Journal,* 8, 1665, 1989b.

Yoshida, M. and Precht, W., Monosynaptic inhibition of neurons of the substantia nigra by caudato-nigral fibers, *Brain Res.,* 32, 225, 1971.

Yung, W. H., Hausser, M. A. and Jack, J. J. B., Electrophysiology of dopaminergic and non-dopaminergic neurones of the guinea-pig substantia nigra pars compacta in vitro, *J. Physiol.,* 436, 643, 1991.

Zhang, J.-H., Sato, M. and Tohyama, M., Region-specific expression of the mRNAs encoding β subunits ($\beta1$, $\beta2$, and $\beta3$) of $GABA_A$ receptor in the rat brain, *J. Comp. Neurol.,* 303, 637, 1991.

CHAPTER 6

A STRATEGY FOR MEASURING NEUROTRANSMITTER
INTERACTIONS *IN VIVO* WITH POSITRON EMISSION
TOMOGRAPHY (PET): NEUROPSYCHIATRIC
IMPLICATIONS

Gwenn S. Smith, Stephen L. Dewey, and Jonathan D. Brodie

I. INTRODUCTION

The understanding of the major psychiatric disorders in a modern scientific framework has lagged behind other medical disciplines, since the theoretical constructs were derived from non-neurochemical measures, primarily the observation and interpretation of complex behaviors interpreted within a psychoanalytic context which hypotheses were difficult to substantiate. As there is no tradition of localizing signs and symptoms, based upon objective, neurophysiological criteria, it has not been possible to evaluate hypotheses regarding the previously suggested loci for disorders such as schizophrenia, depression and anxiety states by observing behavior. Over the past twenty years, advancements in radiotracer chemistry, instrumentation and image processing have resulted in the development and refinement of non-invasive, neuroimaging techniques to measure brain function in the living non-human primate and human (e.g. Positron Emission Tomography [PET] and Single Photon Emission Computed Tomography [(SPECT]). Following from the notion that behavior is a final common path for a complex series of neurochemical and neurophysiological processes, early PET investigations emphasized the use of tracers for glucose metabolism and blood flow which reflect those same processes as biochemical final paths. In the study of the neuropsychiatric disease in general, and more specifically the psychotic disorders, this emphasis led to cross-sectional comparisons of psychiatric populations with control groups conditions of rest, cognitive activation or treatment intervention [Farkas et al., 1980; Farkas et al., 1984; Buchsbaum et al., 1982; Wolkin et al., 1985; Volkow et al., 1986; Tamminga et al., 1986; Gur et al., 1987; Berman et al., 1992; Ganguli et al., 1992; Bartlett et al., 1991; Buchsbaum et al., 1992]. Although metabolic imaging studies appeared to distinguish schizophrenic subjects from normals on the basis of functional organization, the more recent development of positron labeled neurotransmitter receptor ligands [Fowler et al., 1983; Mintun et al., 1984; Wong et al., 1984; Fowler et al., 1990] provided an essential tool to understand potentially the neurochemical basis for these functional differences and effects of treatment.

0-8493-4780-7/96/$0.00+$.50
© 1996 by CRC Press, Inc.

As will be elucidated in the following sections, PET has allowed us to measure the delivery of drugs to the brain and to other organs and, most importantly, to measure the activity of therapeutic agents. Alterations in drug activity can be assessed within a single neurotransmitter system or between functionally-linked systems. Of the currently available neuroimaging and neurochemical methods, this unique application of PET represents the most direct, non-invasive and quantitative method of measuring neurotransmitter activity in the living brain.

II. DEFINITION OF THE CONCEPTUAL FRAMEWORK

A. Psychiatric Considerations

The original clinical impetus for this research approach was the recognition of schizophrenia as an important clinical management and public health problem. Schizophrenia is the most debilitating of all mental illnesses. Its onset is in early adulthood and it results in a failure of the individual to develop psychologically and to function productively in society. Schizophrenia affects 1% of the population (or 3 million individuals), at an annual cost of $10-20 billion (for a review see [Babigan et al., 1985; Cancro et al., 1985]). Not only is schizophrenia associated with an increased rate of suicide and cocaine use, but also with higher and earlier mortality compared to other psychiatric illnesses. Although antipsychotic medications were discovered over forty years ago, there is no cure for the disease. More than 30% of patients are resistant to treatment with antipsychotic medications and approximately 20-30% develop irreversible side effects. Certain symptoms, such as social withdrawal or intellectual impairment, are not alleviated by antipsychotic drugs. Even though several new agents have been developed, a large percentage of patients are still refractory to all known drugs. In fact, there is no predictive test to determine which of the existing medications will be efficacious and will produce the least side effects for a given patient. The appropriate treatment regimen is determined by trial and error with great suffering and cost to the patient and family. **Given these considerations, we sought to develop an *in vivo* biologic marker to examine disease mechanisms and to predict treatment response, all of which will lead ultimately to innovations in drug development.**

There is a widely accepted notion that the therapeutic effect of neuroleptics in the treatment of schizophrenia evolves over days to weeks or even months [Miller, 1987]. Such a delay in therapeutic efficacy cannot be explained by a delay in central dopamine receptor blockade, since blockade of the striatal D2 receptor is established within hours of neuroleptic administration [Nordstrom et al., 1992; Wong et al., 1984]. Furthermore, relapse following cessation of treatment often does not occur until weeks or even months after neuroleptic plasma levels become undetectable [Johnson et al., 1983]. While abrupt cessation of medication in chronic schizophrenics results in a gradually increasing D_2 receptor availability over at least seven days after the final neuroleptic dose [Smith et al., 1988] and is directly correlated with plasma haloperidol level [Wolkin et al., 1986] studies in our own [Wolkin et al., 1989] and other laboratories [Cambon et al., 1987; Coppens et al., 1991; Farde et al., 1988] have also shown that administration of therapeutic doses of neuroleptics produces equivalent striatal dopamine receptor occupancy in both treatment responsive and non-responsive patients. This indicates that failure to respond is not due to the failure of neuroleptics to occupy the striatal dopamine receptors and further supports the notion that a therapeutic response may be related to the involvement of other neurochemical processes.

However, PET neurotransmitter receptor studies to date have been limited largely to the examination of the striatal D2 dopamine receptor, due to technical limitations, while the importance of meso-cortical/meso-limbic dopamine systems in schizophrenic symptomatology has been stressed [Stevens, 1993; Torrey and Peterson, 1974]. However, several factors hinder the ability to image these dopaminergic pathways directly, including limitations of spatial resolution (which precludes imaging of mesencephalic dopamine cell bodies), the difficulties of measuring extrastriatal (cortical and limbic) dopamine receptors due to the low density of these receptors and the lack of available radiotracers that bind significantly to these receptors. The available PET technology is particularly well suited to developing a different and potentially clinically relevant alternative approach: to measure dopamine modulation of other neurotransmitters with a widespread cortical and limbic distribution that are known to interact with dopamine. Interactions between dopamine and GABA, dopamine and acetylcholine and dopamine and serotonin have been well-documented. In addition, there is evidence of the importance of GABAergic and extra-striatal (cortical and limbic) serotonergic and cholinergic systems in treatment response (particularly of the GABA agonists and atypical neuroleptics) and dysfunction of these neurotransmitter systems in certain aspects of schizophrenic symptomatology [Deutsch et al., 1991; Meltzer et al., 1991; Tandon and Gredon, 1989].

Rather than adopting a conceptual framework based upon measuring the static properties of striatal dopamine receptors in isolation, we have chosen to use PET to measure the dynamic properties of the dopamine system by examining its ability to modulate neurotransmitters to which it is functionally linked and to measure its modulation by other neurotransmitters. Symptomatology and treatment response may be related to other factors, such as the integrity of either extra-striatal dopamine systems or other neurotransmitter systems, functionally-linked to dopamine, shown to be altered in schizophrenia or responsive to neuroleptic administration. In addition to dopamine, the logical neurotransmitter systems for study, based upon their neurophysiologic role and clinical significance, are GABA, serotonin and acetylcholine.

B. Basic Neuroscience Considerations

It is essential to conduct PET studies within a neuroanatomically and neurophysiologically well-defined system to provide a context within which the PET data are interpreted. Ideally, the neurochemical constituents of such a system would include those neurotransmitters implicated in schizophrenia, but also in the other major neuropsychiatric disorders. The extrapyramidal motor system fulfilled these criteria. The anatomy and physiology of the extrapyramidal motor system is well known and disturbances within the system have been implicated in many neuropsychiatric diseases (e.g. Parkinson's disease, Huntington's disease and schizophrenia). The main component of this system, the corpus striatum, is large enough to permit imaging within the resolution of the PET scanner. Importantly, the same interactions that occur in the striatum also occur in cortical and limbic areas that can be imaged with PET, as well. Initial PET studies were conducted to measure modulation of the extra-pyramidal motor system by neurotransmitters intrinsic (dopamine, acetylcholine, GABA) and extrinsic (serotonin) to this system, as will be described in the next section.

To review briefly, the cell bodies of origin of **cortical and limbic** dopaminergic, serotonergic and cholinergic systems are located in the mesencephalon (substantia nigra and ventral tegmental area for the dopamine system, raphe nuclei for the serotonin system) and basal forebrain (the medial septum, vertical limb of the diagonal band nucleus and the nucleus basalis of Meynert for the cholinergic system [Dahlstrom and Fuxe, 1964; Lindvall and Bjorklund, 1974; Steinbusch, 1991; Mesulam et al., 1983]). Striatal and pallidal GABAergic neurons project onto dopaminergic neurons of the substantia nigra and ventral tegmental area [Kubota et al., 1987; Smith and Bolam, 1990]. Neuroanatomic methods (including electron microscopic studies) have demonstrated interconnections between the cell bodies of origin for dopaminergic, cholinergic and serotonergic systems, which suggest that neurotransmitter interactions may occur at the level of the cell bodies [Fibiger and Miller, 1977; Jones and Cuello, 1989; Zaborsky, 1991]. Diffuse cortical projections have been characterized for the cholinergic, serotonergic and GABAergic systems, whereas the dopaminergic projections have a more restrictive localization (dorsolateral and orbital prefrontal cortex, premotor and motor cortex, superior and inferior temporal gyri). Consistent with these projections, high to intermediate levels of serotonergic and cholinergic receptors have been described, throughout the cerebral cortex, amygdala and hippocampus [Cortes et al., 1984; Hoyer et al., 1986] .$GABA_A$, benzodiazepine and $GABA_B$ receptors are diffusely distributed throughout cortical and limbic areas and relatively high densities are found in substantia nigra, amygdala and cerebellum [Bowery et al., 1987,

Cortes et al., 1987; Zezula et al., 1988]. Similarities in the cortical laminar organization of these receptors suggest that these systems may also interact at the level of the terminal projection fields [Luabeya et al., 1984]. A summary of interactions between the neurotransmitters of interest is provided in Table 1.

TABLE 1: NEUROTRANSMITTER INTERACTIONS

	GABA	ACh	DA	5-HT
GABA		+	--	--
ACh	+		--	--
DA	--	--		--

note: + = increase in activity (decrease in receptor binding)
       -- = decrease in activity (increase in receptor binding)
       ACh = acetylcholine
       DA = dopamine

III. NEUROTRANSMITTER ACTIVITY AND INTERACTIONS MEASURED WITH PET: EXPERIMENTAL DATA

A. Methodological Considerations

The ability to measure neurotransmitter activity *in vivo* with PET was made possible by advancements in radiotracer chemistry. At present, radiotracers are available for a wide variety of neurotransmitter receptor and reuptake sites and more recently even for second messengers [Imahori et al., 1984]. There are several phases of testing, both preclinical and clinical, to determine whether a radiotracer has the properties suitable for imaging of the neurotransmitter system of interest in non-human and human primates. For the purposes of imaging, it is important that the radiotracer demonstrates suitable kinetic properties. In addition, the radiotracer should demonstrate a high ratio of specific to non-specific binding in the target regions, the regional pattern of binding is selective for a given receptor system, and the regional pattern of binding corresponds with the known receptor localization demonstrated by autoradiographic studies. The radiotracer must demonstrate low test-retest variability, as a measurement of the endogenous variance of these systems in the unperturbed state. Finally, radiotracer binding must be sensitive to alterations in the endogenous competitor induced by pharmacological challenge. This means that the radiotracer should be readily reversible and of moderate affinity for the receptor of interest. When possible, within a given neurotransmitter system, the challenge studies are done with drugs that act by different mechanisms of action. The research strategy described is predicated upon the observation that, *in vivo*, the radiotracer competes with the endogenous neurotransmitter for binding to the receptor, such that if there is a pharmacologic-induced increase in neurotransmitter activity, this will result in a decrease in radiotracer binding due to an increase in the competition for receptor binding. In contrast, if there is a pharmacologic-induced decrease in neurotransmitter activity, this will result in an increase in radiotracer binding due to a decrease in the competition for receptor binding. Thus, one difference between in vitro and *in vivo* receptor binding studies is the ability to capitalize on the competition between the radiotracer and the endogenous neurotransmitter. The other difference is that *in vitro*, the affinity of a ligand for a receptor is the major determinant of ligand binding; *in vivo* the halftime of the tracer in tissue is the major determinant [Logan et al., 1991]. This became apparent after demonstrating that the binding of radiotracers that possessed high (^{18}F-NMSP) and low (^{11}C-raclopride) affinities for the striatal D2 receptor could be reduced in both cases by a pharmacologic increase in dopamine activity by amphetamine administration [Dewey et al., 1991, 1993].

The next sections will address the specific baboon and human studies performed to demonstrate the feasibility of the experimental approach. The first series of studies focused on the modulation of striatal dopamine activity, the main component of the extrapyramidal motor system, by GABA, acetylcholine and serotonin, the primary neurotransmitters known to modulate dopamine activity in this pathway. Next, the studies performed using the muscarinic cholinergic radiotracer ^{11}C-benztropine will be described. Given the widespread distribution of muscarinic receptors in cortical and limbic areas, the use of this tracer has enabled us to measure interactions in these areas, which may be more sensitive to disease processes and treatment interventions.

B. Modulation of Dopamine Activity by GABA, Acetylcholine and Serotonin

B.1 The Dopamine Radiotracer: Establishing Test-Retest Variability and Sensitivity of the Radiotracer to Alterations in Dopamine Activity

Initial studies with the radiotracer ^{18}F-n-methyl-spiroperidol (^{18}F-NMSP, a positron labeled dopamine 2 and serotonin2 ligand, structurally related to the neuroleptic haloperidol) were undertaken to establish test-retest variability of the radiotracer and to demonstrate the sensitivity of radiotracer binding to alterations in dopamine activity. The test-retest variability of ^{18}F-NMSP was less than 5% in the adult female baboons and averaged 9% in the human subject studies. Test-retest variability of this magnitude has been demonstrated for ^{11}C-raclopride (a more selective, reversible D2 specific ligand) in baboons and humans [Dewey et al., 1992; Volkow et al., 1993].

Interaction studies are contingent upon the ability to demonstrate alterations in ligand binding (as an index of receptor availability) after pharmacologically-induced changes in neurotransmitter activity. In fact, differences in neurotransmitter activity (as it would affect the binding of ligands with different affinities for dopamine) may account for inconsistent reports of increased striatal dopamine receptor number in naive schizophrenics, as compared with normal controls [Andreassen et al., 1988; Seeman et al., 1990; Wong et al., 1986, Farde et al., 1990]. PET studies were performed to measure change in ^{18}F-NMSP binding after increase of dopamine release by acute administration of amphetamine and demonstrated a reduction in binding in striatum (specific binding), not in cerebellum (non-specific binding), consistent with an increase in competition for binding to the receptor site [Dewey et al., 1991]. The metabolite corrected plasma input function and the rate of metabolism of the radiotracer were unaltered by the drug intervention. A variation in response was noted across the baboons and has been reported also using other neurochemical methods and behavioral measures [Kuzenski and Segal, 1989]. Therefore, changes in neurotransmitter activity can be detected by alterations in receptor availability with PET, even when using a ligand with high affinity for the dopamine receptor, such as ^{18}F-NMSP [Logan et al., 1991].

Using the same d-amphetamine challenge paradigm as with ^{18}F-NMSP, a reduction in ^{11}C-raclopride binding was observed in the striatum, but no change was observed in cerebellum. The K1 transport constant, the metabolite corrected plasma input function and the rate of metabolism of the radiotracer were unaltered. It is important to note that consistent with the findings of Seeman et al. (1989) within the same baboon, the effect of d-amphetamine on ^{11}C-raclopride binding was greater than on ^{18}F-NMSP binding, due to the relatively greater affinity of ^{18}F-NMSP for the D2 receptor. The sensitivity of ^{11}C-raclopride binding to alterations in dopamine has been demonstrated also using other pharmacologic agents that act by different mechanisms of action (GBR 12909, cocaine, methylphenidate [Dewey et al., 1993]). **Therefore, the sensitivity of ligand binding to alterations in dopamine has been demonstrated with two ligands, each possessing different affinities for the dopamine receptor. In the case of ^{11}C-raclopride the results were confirmed with drugs that increase synaptic availability of dopamine by different mechanisms of**

action, by blocking reuptake (GBR 12909, cocaine, methylphenidate) or promoting cytosolic release (d-amphetamine). In the human, the sensitivity of ^{11}C-raclopride binding to alterations in dopamine activity has been demonstrated recently [Volkow et al., 1994] using the dopamine reuptake inhibitor methylphenidate. Variability in response across subjects was observed, which was significantly correlated with baseline clinical ratings of anxiety and depression.

B.2 GABA-Dopamine Interaction

The GABAergic system was selected for study due to importance in the nervous system as the primary inhibitory neurotransmitter in the brain (its concentration is 200-1000 times greater than dopamine, acetylcholine and serotonin [McGeer and McGeer, 1989] and its inhibition of striatal dopamine is well documented [Bunney and Aghajanian, 1976]. In addition to the neurophysiologic significance of GABA, previous studies have demonstrated GABAergic abnormalities in schizophrenic brain tissue and CSF [Benes et al., 1992, Perry et al., 1979; Thaker et al., 1987]. GABA potentiation may represent a clinically effective alternative route for decreasing dopamine activity (by administration of either a benzodiazepine or GABA transaminase inhibitor alone or with a lower neuroleptic dose [Stahl et al., 1985, Tamminga et al., 1983; Wolkowitz and Pickar, 1991]. A measure to assess the GABA-dopamine interaction *in vivo* could be used to determine whether this pathway was altered in schizophrenia and whether this would represent an efficacious treatment alternative for a given patient.

Administration of either the anticonvulsant gamma vinyl-GABA (GVG), or a widely prescribed benzodiazepine agonist, lorazepam resulted in an increase in ^{11}C-raclopride binding in the striatum (specific binding), but not in the cerebellum (non-specific binding) [Dewey et al., 1992]. In both cases, these increases significantly exceeded the test/retest variability of this radiotracer and the metabolite corrected plasma input function and the rate of metabolism of the tracer was not altered by drug administration. This study represented the first demonstration of an increase in radiotracer binding following a drug challenge with PET [Dewey et al., 1992].

B.3 Acetylcholine-Dopamine Interaction

Studies have been conducted in baboons and in human subjects to measure the effect of central cholinergic blockade on striatal dopamine receptor availability [Dewey et al., 1990; 1993]. Clinically, an imbalance between acetylcholine and dopamine in the striatum is manifest in terms of movement abnormalities (e.g. the extrapyramidal side effects of neuroleptic treatment, akinesia and rigidity). Therefore, the development of a measure to assess this interaction could be important in determining whether a given patient (especially such vulnerable populations as the elderly and females) would be predisposed to these side effects observed as a result of neuroleptic or antidepressant medication.   According to the connectivity of the extrapyramidal motor system, cholinergic blockade would increase the synaptic availability of striatal dopamine by inhibiting excitatory input to GABA neurons [Ferkany and Enna, 1980]. The importance of cholinergic input to GABA neurons was supported by the finding that the increase in dopamine receptor binding (of [3H]-spiroperidol) induced by GABA agonist (THIP) administration could be blocked by the muscarinic receptor antagonist, atropine. Atropine had no effect when administered by itself [Ferkany and Enna, 1980].

Using the radiotracer ^{18}F-NMSP, studies in both baboons and normal male human subjects,   administration of the widely used anticholinergic medication benztropine (Cogentin) resulted in a decrease in striatal binding of ^{18}F-NMSP, without change in the cerebellum [Dewey et al., 1990]. The metabolite corrected plasma input function and the rate of ^{18}F-NMSP metabolism in plasma was not altered. This study was replicated using both a radiotracer with a higher specificity and lower affinity for the D_2 receptor (^{11}C-raclopride) and a muscarinic antagonist which has a more selective mechanism of action (scopolamine [Dewey et al., 1993]). In normal, male control subjects, ^{11}C-raclopride binding was decreased in the striatum, but not in the cerebellum, in excess of the test-retest variability of the ligand (5%). The metabolite corrected plasma input function and the rate of metabolism of the tracer was not altered by drug administration. These results are consistent with the physiology of the nigro-striatal dopamine system and with the ^{18}F-NMSP/benztropine data.

B.4 Serotonin-Dopamine Interaction

The importance of the serotonin-dopamine interaction and the role of the serotonin$_2$ (5-HT$_2$) receptor, in particular, is highlighted by its relatively greater density and neurophysiologic role in both cerebral cortex and striatum and abnormalities of this particular interaction and receptor subtype in schizophrenia and in the mechanism of action of atypical neuroleptics and antidepressant medications [Peroutka et al., 1989; Pazos et al., 1987; Palacios et al., 1990; Arora and Meltzer, 1991; Deutsch et al., 1991; Meltzer, 1991].Initial studies were performed in baboons using the antagonist altanserin, which has a greater selectivity for serotonergic receptors versus noradrenergic and dopaminergic receptors, compared with other available antagonists. ^{11}C-raclopride binding was reduced in the striatum, in excess of the test-retest variability of the ligand [Dewey et al., 1995]. No alterations were observed in the cerebellum. The metabolite corrected plasma input function and the rate of metabolism of the tracer were not altered by drug administration. Consistent with its known physiologic role, serotonin blockade disinhibited the striatal dopaminergic neuron thereby increasing dopamine activity and decreasing ^{11}C-raclopride binding.

Subsequent studies have been performed in baboons using the selective serotonin reuptake inhibitor, citalopram [Dewey et al., 1995]. In this case, ^{11}C-raclopride binding was increased in the striatum, but not in the cerebellum, in excess of the test-retest variability of the ligand. Finally, studies have been undertaken in human subjects, with the serotonin releaser and reuptake inhibitor, fenfluramine [Smith et al., 1994]. Fenfluramine was chosen as it is the most selective and widely used challenge drug for the serotonin systems. There is a considerable amount of data to demonstrate that fenfluramine administration produces a decrease in prolactin release, which may occur through interactions with a dopaminergic mechanism [McBride et al., 1990]. The ^{11}C-raclopride scans were performed at the point when maximal plasma drug and prolactin levels were observed. Subjects showed a decrease in D2 receptor binding, in contrast to the baboon findings. It is important to note that of all the interactions studied thus far, serotonin's inhibition of dopamine is the most controversial. While some *in vivo* microdialysis studies are supportive of the direction of the interaction observed in the PET studies [Dewey et al., 1995; Meltzer et al., 1993], others are not [Benloucif and Galloway, 1990]. Basic neuroscience studies indicate that that the status of the dopamine system at the time of challenge (i.e. stimulated versus resting) may determine the nature of the interaction [Benloucif and Galloway, 1990]. Subsequent studies will be designed to address this possibility. In addition, studies will be performed with a new serotonin (5HT-2) radiotracer to measure the modulation of serotonin by dopamine in cortical and limbic areas [Tan et al., 1994].

C.   Modulation of Acetylcholine Activity by Dopamine, Serotonin and GABA

C.1   Development of ^{11}C-benztropine as a Radiotracer for the Muscarinic Cholinergic Receptor

[11]C-benztropine has been developed as a ligand for the muscarinic cholinergic receptor [Dewey et al., 1990]. Benztropine was chosen from among the available cholinergic antagonists because (1) it has a long biological half life and remains intact in the systemic circulation, an advantage compared to other muscarinic ligands such as ^{11}C-scopolamine, which is rapidly metabolized in plasma [Frey et al., 1992] and (2) its clinical use (as an anticholinergic to relieve the extrapyramidal side effects of neuroleptic medication) makes it feasible to administer intravenously in unlabelled form for pharmacological blockade. The ligands ^{11}C-tropanyl benzilate and ^{11}C-N-methyl-4-piperidyl benzilate have been developed also as radiotracers for the muscarinic receptor [Koeppe et al., 1992; Mulholland et al., 1992]. Striatal binding values are obtained with these ligands that are

consistent with the ^{11}C-benztropine results, using a distribution volume analysis.

In the baboon and human, incorporation of tracer was greatest in striatum and occipital cortex, consistent with post-mortem autoradiography data in human brain [Cortes et al., 1984, 1987]. Uptake was also observed in frontal, temporal and parietal cortices, thalamus and hippocampus. Test-retest variability ranged from 0 to 13%, across brain regions, with an average of 10% or less across all regions. Administration of a clinically effective dose of unlabelled benztropine or scopolamine reduced binding in striatum, cortex and thalamus and altered binding to a lesser extent in cerebellum, but did not alter the metabolite corrected plasma input function or the rate of metabolism of ^{11}C-benztropine. Since pharmacologic doses of benztropine have been shown to inhibit dopamine reuptake {Coyle and Snyder, 1969], studies were undertaken to determine whether ^{11}C-benztropine binds to a significant extent to the dopamine reuptake site. Baboons were pretreated with the potent dopamine reuptake inhibitors nomifensine, cocaine or GBR 12909 and in all cases tracer incorporation was not changed in excess of the magnitude of test-retest variability; changes were particularly small in the striatum, which contains the highest concentration of dopamine reuptake sites [Fowler et al., 1987]. In these studies, the metabolite corrected plasma input function and the rate of metabolism of the tracer were not altered by drug administration. Studies using three different dopamine reuptake inhibitors demonstrated that ^{11}C-benztropine does not bind to a significant extent to the dopamine reuptake site. It is important to note that labeled benztropine is administered in high specific activity (low mass), such that the drug is at extremely low concentration at the binding site. In summary, ^{11}C-benztropine is a relatively selective ligand for the muscarinic cholinergic receptor, for which good test-retest reliability is obtained. The regional distribution of ^{11}C-benztropine binding is consistent with the known distribution of muscarinic receptors, its binding is not significantly altered by dopamine reuptake inhibitors, indicating that it is not binding significantly to these sites. While there are limitations with all of the available cholinergic ligands (e.g. difficulties in quantitation), newer ligands are in various stages of development.

C.2   Dopamine-Acetylcholine   Interaction

To measure the effect of central dopaminergic blockade on muscarinic receptor availability, ^{11}C-benztropine binding was measured after administration of NMSP (2.8 ug/kg, IV), in the baboon [Dewey et al., 1990]. Binding was reduced in striatum, frontal cortex , thalamus and cerebellum. The rate of metabolism of ^{11}C-benztropine in plasma and the metabolite corrected plasma input function were unaltered. As predicted from the neuroanatomic connections, blockade of dopamine receptors may have increased acetylcholine release by disinhibiting cholinergic interneurons [Bloom et al., 1965; Connor et al., 1970]. The possibility that binding is reduced due to the competition of increased dopamine with ^{11}C-benztropine for the dopamine reuptake site is less likely given that nomifensine, cocaine and GBR 12909 administration did not alter ^{11}C-benztropine binding. A subsequent study (using the serotonin antagonist altanserin) was conducted which addresses the contribution of NMSP binding to the serotonin (5-HT2) receptor. This would also increase acetylcholine release and decrease receptor availability.

C.3 GABA-Acetylcholine Interaction

The effects of GVG on ^{11}C-benztropine binding were assessed to measure the GABA-acetylcholine interaction {Dewey et al., 1993]. As GABA is known to inhibit dopamine activity, it was predicted that in turn, the cholinergic neuron would be disinhibited, releasing more acetylcholine. ^{11}C-benztropine binding was decreased in a regionally-specific manner. The greatest effects were observed in striatum and frontal cortex. The effects in thalamus and cerebellum were much smaller and did not exceed the test-retest variability of the ligand. The larger effect in striatum, relative to other regions, and the lack of an effect in thalamus is consistent with neurophysiologic data [Cosi and Wood, 1988]. The metabolite corrected plasma input function and the rate of metabolism of the tracer were not altered by drug administration.   These results, considered with the GVG/^{11}C-raclopride data, demonstrate that in the same time frame, administration of a drug with a primary effect on the GABA system produced differential effects on two functionally-linked neurotransmitters and these effects were consistent with the known physiology of these systems.

C.4 Serotonin-Acetylcholine Interaction

The effects of serotonergic (5-HT2) receptor blockade on the binding of ^{11}C-benztropine were examined [Dewey et al., 1993]. Administration of altanserin produced a regionally-specific decrease in ^{11}C-benztropine binding consistent with the known anatomy of these two systems. Unlike the GVG studies, decreases in labeled benztropine binding were noted only in the striatum. These decreases significantly exceeded the test/retest variability of this radiotracer. The metabolite corrected plasma input function and the rate of metabolism of the tracer were not altered by drug administration. In regard to the ^{11}C-benztropine/ NMSP results, the effects observed in the cortex are most likely due to binding to dopamine, rather than serotonin receptors. Regional differences in ^{11}C benztropine distribution volumes between these studies and those using GVG demonstrate that PET can be used to identify specific regional effects on radiotracer binding following pharmacological challenge.

D. Summary

The studies described in this section have demonstrated that PET can be used to measure neurotransmitter activity within single systems and to measure neurotransmitter interactions within functionally-linked systems. The feasibility of using this experimental approach in human subjects is supported by the studies performed thus far. We have shown that the modulation of dopamine activity by GABA, acetylcholine and serotonin and the modulation of acetylcholine activity by dopamine, GABA and serotonin can be measured *in vivo* with PET. In order to provide independent validation of these findings, *in vivo* microdialysis studies were conducted to monitor changes in extracellular dopamine as a function of the same pharmacologic challenges used in the PET studies. In all cases, the microdialysis data were consistent with the PET data [Dewey et al., 1995].

IV. IMPLICATIONS OF THE EXPERIMENTAL APPROACH

The extension of the experimental approach to human subjects and clinical populations will permit us to obtain information regarding the mechanisms underlying basic brain function, disease processes and treatment response that could not be obtained previously with other available methods. The implications of these studies can be classified into three major areas. The use of PET with radiotracers to study the interaction of neurotransmitter systems *in vivo* will allow us: **(1) To test directly whether observations made using basic neuroscience methodologies in rodents and non-human primates are valid in the living human brain**. One example is an active area of investigation in the field of clinical neuroscience, the significance of neural circuits involving prefrontal cortex and hippocampus in cognition and in the cognitive deficits observed in neuropsychiatric illness [Goldman-Rakic, 1987; Weinberger, 1993]. This approach will provide unique information regarding neurochemical modulation within these same brain regions.

(2) To test new hypotheses about the etiology of neuropsychiatric disease. Previous studies to measure neurochemical alterations in psychiatric illness have examined brain tissue at autopsy, or biochemical markers in plasma, or cerebrospinal fluid, all of which are indirect measures of central neurochemical activity *in vivo*. As a result, many neuropsychiatric disorders (e.g. schizophrenia, Alzheimer's disease, Parkinson's disease) have been attributed classically to deficits within single neurotransmitter systems. Therapies directed at reversing these putative deficits have not been efficacious consistently and have not stopped disease progression. The use of PET to study neurotransmitter interactions will permit us to assess the functional integrity of these neurotransmitter systems, specifically the ability of one system to modulate another anatomically and functionally linked system. This will allow us to test alternative hypotheses of the etiology of neuropsychiatric disease, and thus will provide important information regarding the design of more effective treatments. Other potential clinical applications include the assessment of variability in symptomatology. This would permit us to assess the neurochemical substrates of the clinical heterogeneity observed in schizophrenia and other neuropsychiatric diseases [Andreassen and Carpenter, 1993].

(3) **To develop a predictive marker of treatment response in schizophrenia, depression and other neuropsychiatric diseases, based upon measures of neurochemical plasticity.** In the case of schizophrenia, we have hypothesized that the inability to respond to treatment is predicted by a lack of neurochemical plasticity, namely a failure of dopamine to modulate cortical and limbic cholinergic systems. This methodology could be extended further to assess the mechanisms of drug action and to assess the impact of treatment. If dopamine modulation is found to be critical in treatment response, new medications in development could be screened for their ability to alter dopamine modulation and appropriate dosing regimens could be developed. This application of PET could contribute to the development of novel and more rational treatments for schizophrenia and other diseases and shorten the long process of drug discovery.

V. SUMMARY AND FUTURE DIRECTIONS

The studies described in this chapter demonstrate the use of PET to measure dynamic neurochemical responses to pharmacologically specific interventions by relating downstream changes in receptor occupancy to upstream neurotransmitter *activity* in functionally coupled neurotransmitter systems *in vivo*. The data obtained are remarkably consistent with the basic neuroscience studies that initially described these interactions using neurochemical and neurophysiologic methods Not only are the immediate consequences of drug activity discernible with the appropriate labeled agents, as shown by the change in elaboration of a neurotransmitter which is functionally coupled to the target neuron, but the coupling of an initial system to another can be followed at least two synapses away as shown by the effect of GABA facilitation on cholinergic binding and by final functional connection as reflected by measures of regional glucose utilization.

The data with a multiplicity of systems supports the notion of relatively rapid response to a pharmacological intervention at the synaptic system level and a relatively delayed response at the level of energy utilization. The temporal adaptation is also observed clinically in human subjects, as evidenced by the data showing a delayed glucose metabolic response to haloperidol challenge in normal human subjects [Bartlett et al.,

1994]. The most significant observation is the dynamic aspect of the interactions. All of the expected effects of the initial response to a given pharmacological perturbation were observed. Thus, a variety of agents which increase synaptic dopamine caused a decrease in binding of labeled raclopride. Two different agents which facilitated GABAergic transmission caused the expected decrease in synaptic DA and an increase in raclopride binding. It is particularly noteworthy that GVG, a suicide enzyme inhibitor, produced the expected changes at the cholinergic receptor in the same time frame as changes were seen at the striatal D2 receptors.

The studies described illustrate how our experimental approach can be used to address the pathophysiology and pharmacotherapy of neuropsychiatric diseases that have been classically attributed to deficits in a single neurotransmitter system. While a disease may begin with a single neurotransmitter deficit, the progression of a disease, such as Kraepelinian schizophrenia, undoubtedly extends to other neurotransmitters functionally linked to the initial locus. This functional coupling suggests that treatment strategies aimed at altering activity in other systems linked to dopamine, such as the serotonergic, cholinergic and GABAergic systems, could be efficacious when loss of plasticity of the target dopamine receptors can be demonstrated. Functional coupling is certainly consistent with observations that a multiplicity of systems may be targeted in the treatment of depressive syndromes and the psychiatric symptoms in Alzheimer's disease, as well.

Studies are in progress to use these experimental strategies to assess the effects of chronic neuroleptic treatment in baboons. This will allow us to measure the effects of chronic treatment on neurotransmitter interactions in the absence of a disease process, which cannot be done in humans since normal subjects cannot be given neuroleptics on a chronic basis. Human studies are also in progress to measure the ability of dopamine to modulate cholinergic and serotonergic activity and the ability of glutamate to modulate dopamine activity, as a means of measuring cortical regulation of striatal dopamine. Having developed the experimental parameters in control subjects, these studies will be conducted in schizophrenic patients classified as treatment responders and non-responders. In addition, studies are being conducted in human subjects with the more clinically available and cost-effective imaging method, single photon emission computed tomography (SPECT), to develop similar research strategies. The work of Kung and colleagues [e.g. Kung et al., 1990] has resulted in the development of radiotracers for the dopamine and serotonin receptors and reuptake sites. In addition, the sensitivity of the SPECT D2 dopamine ligand ^{123}I-IBZM to alterations in endogenous dopamine in non-human primates has been demonstrated [Innis et al., 1992]. Their results invite the application of SPECT to neurotransmitter interaction studies and hence an application in clinical rather than research settings. Although we may be unduly optimistic, it is our hope that this would result in the use of this technology as a clinical instrument for the development of treatment regimes tailored for an individual patient in which the therapeutic effects and side effects can be properly balanced.

REFERENCES

Andreasen, N., Carpenter, W. Diagnosis and classification of schizophrenia. Schiz Bull . 1993; 19: 199-214.

Andreasen, N., Carson, R., Diksic, M., Evans, A., Farde, L., Gjedde, A., Hakim, A., Lal, S., Nair, N., Sedvall, G., Tune, L., Wong, D. Workshop on schizophrenia, PET and dopamine D2 receptors in human neostriatum. Schiz Bull. 1988; 14: 471-484.

Arora, R., Meltzer, H. Serotonin (5HT-2) receptor binding in the frontal cortex of schizophrenic patients. J Neur Trans. 1991; 85: 19-29.

Babigian, H. Schizophrenia: Epidemeology. Comprehensive Textbook of Psychiatry, Kaplan, H, Sadock, B. (eds.), Williams and Wilkins, Baltimore, M.D., 1985; pp. 643-650.

Bartlett, EJ; Brodie JD; Simkowitz P; Dewey S; Rusinek H; Wolf AP; Fowler JS; Volkow ND; Smith G; Wolkin A; Cancro R. Effects of haloperidol challenge on regional cerebral glucose utilization in normal human subjects. Am J Psychiat. 1994; 151: 681-686.

Bartlett,EJ, Wolkin A, Brodie JD et al. Importance of pharmacologic control in PET studies: effects of thiothixene and haloperidol on cerebral glucose utilization in chronic schizophrenia. Psychiatry Res: Neuroimag 1991;40:115-124.

Benloucif, S., Galloway, M. Facilitation of dopamine release *in vivo* by serotonin agonists: studies with microdialysis. Eur J Pharm. 1991; 200: 1-8.

Benes, F., Vincent, S., Alsterberg, G., Bird, E., San Giovanni, J. Increased GABAa receptor binding in superficial layers of cingulate cortex in schizophrenics. J Neurosci. 1992; 12; 924-929.

Berman, K., Torrey, F., Daniel, D., Weinberger, D. Regional cerebral blood flow in monozygotic twins discordant and concordant for schizophrenia. Arch Gen Psychiat.1992; 49: 927-934.

Bloom, F., Costa, E., Salmoiraghi, G. Anaesthesia and the responsiveness of individual neurons of the caudate nucleus of the cat to acetylcholine, norepinephrine and dopamine administered by microelectrodes. J Pharm Exper Therap. 1965; 50: 244-252.

Bowery, N., Hudson, A., Price, G. GABAa and GABAb receptor site distribution in the rat central nervous system. Neurosci. 1987; 20: 365-383.

Buchsbaum, MS, Ingvar DH, Kessler R, Waters R, Cappelletti J, VanKammen D, King AC, Johnson JC, Manning RF, Flynn RW, Mann LS, Bunney WE, Sokoloff L Cerebral glucography with positron tomography. Arch Gen Psychiat. 1982;41, 1159-1168.

Buchsbaum, MS, Potkin SG, Marshall JF et al. Effects of clozapine and thiothixene on glucose metabolic rate in schizophrenia. Neuropsychopharm. 1992; 6:155-163.

Bunney, B., Agahajanian, G. Dopaminergic influence in the basal ganglia: evidence for nigro-striatal feedback regulation. In The Basal Ganglia (M.Yahr, ed.). New York, Raven Press, 1976; pp. 249-267.

Cancro, R. History and overview of schizophrenia. Comprehensive Textbook of Psychiatry, Kaplan, H, Sadock, B. (eds.), Williams and Wilkins, Baltimore, M.D., 1985 pp. 631-642.

Cambon, H., Baron, J-C., Boulenger, J., et al. *In vivo* assay for neuroleptic receptor binding in the striatum. Brit J Psychiat. 1987; 151: 824-830.

Connor, C. Caudate nucleus neurons: correlation of the effects of substantia nigra stimulation with iontophoretic dopamine. J Physiol. 1970; 208: 691.

Coppens, H., Slooff, C., Paans, A., Wiegman, T., Vaalberg, W., Korf, J. High central D2-dopamine receptor occupancy as assessed with positron emission tomography in medicated but therapy resistant schizophrenic patients. Biol Psychiat, 1991; 29:629-634.

Cosi, C., Wood, P. Lack of GABAergic modulation of acetylcholine turnover in the rat thalamus. Neurosci Lett. 1988; 87: 293-296.

Cortes, R., Probst, A., Palacios, J. Quantitative light microscopic autoradiographic localization of cholinergic muscarinic receptors in human brain: brainstem. Neurosci, 1984; 12: 1003-1026.

Cortes, R., Probst, A., Palacios, J. Quantitative light microscopic autoradiographic localization of cholinergic muscarinic receptors in human brain: forebrain. Neurosci. 1987; 20: 65-107.

Coyle, J.T., Snyder, S.H. Antiparkinsonian drugs: inhibition of dopamine uptake in the corpus striatum as a possible mechanism of action. Science. 1969; 166:899-901.

Dahlstrom, A., Fuxe, D. Evidence for the existence of monoamine containing neurons in the central nervous system. Acta Physiol Scand. 1984; 232: 1-55.

Deutsch, A., Moghaddam, B., Innis, R., Krystal, J., Aghajanian, G., Bunney, B., Charney, D. Mechanisms of action of atypical antipsychotic drugs. Schiz Res. 1991; 4: 121-156.

Dewey, S.L., Logan, J., Wolf, A., Brodie, J., Angrist, B., Fowler, J., Volkow, N. Amphetamine induced decreases in ^{18}F-N-methylspiroperidol binding in the baboon brain using Positon Emission Tomography (PET). Synapse. 1991; 7:324-327.

Dewey, SL, Smith GS, Logan J, Brodie JD. Modulation of Central Cholinergic Activity by GABA and Serotonin: PET Studies with ^{11}C-Benztropine in Primates. Neuropsychopharm. 1993; 8: 371-376.

Dewey, S.L., Macgregor, R., Brodie, J., Bendriem, B., King, P., Volkow, N., Schleyer, D., Fowler, J., Wolf A., Gatley J., Hitzemann, R. Mapping muscarinic receptors in the human and baboon brain using N-^{11}C-methyl-benztropine. Synapse. 1990; 5: 213-233.

Dewey, S.L., Brodie, J., Fowler, J., MacGregor, R., Schleyer, D., King, P., Alexoff, D., Volkow, N., Shuie, C., Wolf, A., Bendriem, B. Positron emission Tomography (PET) studies of the dopaminergic/cholinergic interactions in the baboon brain. Synapse, 1990; 6: 321-327.

Dewey, S., Smith, G., Logan, J., Alexoff, D., King, P., Pappas, N., Brodie, J., Ashby, C. The serotonergic-dopaminergic interaction *in vivo* with Positron Emission Tomography (PET) and microdialysis. J Neurosci. 1995; 15: 821-829.

Dewey, SL; Smith GS; Logan J; Brodie JD; Fowler JS; Wolf AP. Striatal Binding of the PET Ligand ^{11}C-raclopride Is Altered By Drugs That Modify Synaptic Dopamine Levels. Synapse. 1993; 13: 350-356.

Dewey, S., Smith, G., Logan, J., Brodie, J., Yu, D., Ferrieri, R., King, P., MacGregor, R., Martin, T., Wolf, A., Volkow, N., Fowler, J. GABAergic inhibition of endogenous dopamine release measured *in vivo* with ^{11}C-raclopride and Positron Emission Tomography. J Neurosci. 1992; 12: 3773-3780.

Dewey, SL, Smith G, Logan J, Simkowitz P, Brodie JD, Fowler JS, Volkow N, Wolf AP Effects of central cholinergic blockade on striatal dopamine release measured with positron emission tomography (PET) in normal human subjects. Proc Nat Acad Sci. 1993; 90: 11816-11820.

Farde, L, Wiesel F-A, Halldin C, et al Central D2 receptor occupancy in schizophrenic patients treated with antipsychotic drugs. Arch Gen Psychiat. 1988; 45:71-76.

Farde, L., Wiesel, F-A., Stone-Elander, S., Halldin, C., Nordstrom, A., Hall, H., Sedvall, G. D2 dopamine receptors in neuroleptic naive schizophrenic patients. Arch Gen Psychiat. 1990; 47:213-219.

Farkas, T., Reivich, M., Alavi, A., Greenberg, J., Fowler, J., MacGregor, R., Christman, D., Wolf, A. The application of 18-F-2-deoxy-2fluoro-d-glucose and positron emission tomography in the study of psychiatric conditions. In Passonneau, J., Hawkins, R., Lust, W., et al .(eds.): Cerebral Metabolism and Neural Function. Baltimore: Williams and Wilkins, 1980;pp. 403-408.

Farkas, T, Wolf AP, Jaeger J, Brodie JD, deLeon M, Defina P, Christman DR, Fowler JS, MacGregor RR, Goldman A, Yonekura Y: Regional brain metabolism in the study of chronic schizophrenia. Arch Gen Psychiat. 1984; 41:293-300.

Ferkany, J., Enna, S. Interaction between GABA agonists and the cholinergic muscarinic system in rat corpus striatum. Life Sci. 1980; 27: 143-149.

Fibiger, H., Miller, J. An anatomical and electrophysiologic investigation of the serotonergic projection from the dorsal raphe nucleus to the substantia nigra in the rat. Neurosci. 1977; 2: 975-987.

Fowler, J., Arnett, C., Wolf, A., MacGregor, R., Norton, E., Findley, A. ^{11}C-spiroperidol: synthesis, specific activity determination and biodistribution in mice. J Nucl Med 1983;23: 437-445.

Fowler, J., Volkow, N., Wolf, A., Dewey, S.L., Schleyer, D., MacGregor, R., Hitzemann, R., Logan, J., Bendriem, B., Gatley, J., Christman, D. Mapping cocaine binding sites in human and baboon brain in vivo. Synapse. 1989; 4:371-377.

Fowler, J., Wolf, A., Volkow, N., New Directions in Positron Emission Tomography. Part II. Ann Rep Med Chem. 1990; 25: 261-269.

Freed, WJ The therapeutic latency of neuroleptic drugs and nonspecific postjunctional supersensitivity. Schiz Bull. 1988; 14:269-277.

Frey, K., Koeppe, R., Mulholland, G., Jewett, D., Hichwa, R., Ehrenkaufer, R., Carey, J., Wieland, D., Kuhl, D. Agranoff, B. *In vivo* muscarinic cholinergic receptor imaging in human brain with ^{11}C-scopolamine and Positron Emission Tomography, J Cereb Blood Flow Metab. 1992; 12: 147-154.

Friedhoff, AJ A dopamine dependent restitutive system for the maintenance of mental normalcy. Ann New York Acad Sci., 1986; 463:47-52.

Ganguli, R., Mintun, M., Becker, J., Brar, J., Diehl, D., DeLeo, M., Madoff, D., Marditis, A. rCBF during cognitive and physiological stimulation in schizophrenics. Biol Psychiat.1994; 35: 621.

Goldman-Rakic, P. Circuitry of primate prefrontal cortex and regulation of behavior by representional knowledge. In: Plum, F., Mountcastle, V. (eds.) Higher Cortical Function: Handbook of Physiology 5. Washington, DC, American Physiological Society, 1987.

Gur, R., Resnick, S., Gur, R., Alavi, A., Caroff, S., Kushner, M., Reivich, M. Regional brain function in schizophrenia. Arch Gen Psychiat.1987; 44:119-125.

Hoyer, D., Pazos, A., Probst, A., Palacios, J. Serotonin receptors in the human brain II. characterization and autoradiographic localization of 5-HT1C and 5-HT2 recognition sites. Brain Res. 1986; 376:97-107.

Imahori, Y., Ueda, S., Ohmori, Y., Wakita, K., Matsumoto, K. Phosphoinositide turnover imaging linked to muscarinic cholinergic receptors in the central nervous system by positron emission tomography. J Nucl Med. 1993; 34: 1543-1551

Innis, R., Mallison, R., Al-Tikriti, M., Hoffer, M., Sybirska, E., Seibyl, J., Zoghbi, S., Baldwin, R., Laruelle, M., Smith, E., Charney, D., Henninger, G., Elsworth, J., Roth, R. Amphetamine stimulated dopamine release competes *in vivo* for ^{123}I-IBZM binding to the D2 receptor in nonhuman primates. Synapse. 1992; 10; 177-184.

Johnson, DAW, Pasterski G, Ludlow JM et al. The discontinuance of maintenance neuroleptic therapy in chronic schizophrenic patients: drug and social consequences. Acta psychiatr. Scand. 1983; 67:339-352.

Jones, B., Cuello, A Afferents to the basal forebrain cholinergic cell area from pontomesencephalic, catecholamine, serotonin and acetylcholine neurons. Neurosci., 1989; 31: 37-61.

Koeppe, R., Frey, K., Zubieta, J., Fessler, J., Mulholland, G., Kilbourn, M., Mangner, T., Kuhl, D. Tracer kinetic analysis of C-11-N-methyl-4piperidyl benzilate binding to muscarinic cholinergic receptors. J Nucl Med., 1992, 33: 882.

Kubota, Y., Inagaki, S., Kito, S., Wu, J. Dopaminergic axons directly make synapses with GABAergic neurons in the rat neostriatum. Brain Res. 1987; 406: 147-156.

Kuczenski, R., Segal, D. Concomitant characterization of behavioral and striatal neurotransmitter response to amphetamine using *in vivo* microdialysis. J Neurosci. 1989; 9: 2051-65.

Kung, HF, Alavi A, Chang W, Jung, M., Keyes, J., Velchick, M., Billings, J., Pan, S., Noto, R., Rausch, A., Reilly, J. *In vivo* SPECT imaging of CNS D2 dopamine receptors: Initial studies with Iodine-123-IBZM in humans. J Nucl Med. 1990; 573-579.

Lindvall, O., Bjorklund, A. The organization of the ascending catecholamine neuron system in the rat brain. Acta Physiol Scand. 1974; 412, 1-48.

Logan, J., Dewey, S., Wolf, A., Fowler, J., Brodie, J., Angrist, B., Volkow, N., Gatley, S. J. Effects of endogenous dopamine on measures of ^{18}F-N-methylspiroperidol binding in the basal ganglia: comparison of simulations and experimental results from PET studies in baboons. Synapse. 1991; 9: 195-207.

Luabeya, M., Maloteaux, J-M., Laduron, P. Regional and cortical laminar distributions of serotonin, benzodiazepine, muscarinic and dopamine receptor in human brain. J Neurochem. 1984; 43: 1068-1071.

McBride, P., Tierney, H., DeMeo, M., Chen, J., Mann, J. Effects of age and gender on CNS serotonergic responsivity in normal adults. Biol Psychiat. 1990; 27: 1143-1155.

McGeer, P., McGeer, E. Amino acid neurotransmitters. In Basic Neurochemistry. G. Siegel, (ed.), New York, Plenum, 1989; pp. 311-332.

Meltzer, H. The mechanism of action of novel antipsychotic drugs. Schiz Bull. 1991; 17: 263-287.

Meltzer, H., Ickikawa, J., Chai, B. Effect of selective serotonin reuptake inhibitors and tricyclic antidepressants on extracellular dopamine, homovanillic acid and 5-hydroxyindoeacetic acid in rat striatum. Soc Neurosci Abs. 1993; 19 (part 1): 855.

Mesulam, M., Mufson, E., Levey, A., Wainer, B. Cholinergic innervation of the cortex by the basal forebrain. J Comp Neurol. 1983; 214: 170-197.

Mintun, M., Raichle, M., Kilbourn, M., Wooten, G., Welch, M. A quantitative model for the *in vivo* assessment of drug binding sites with Positron Emission Tomography. Ann Neurol. 1984;15: 217-227.

Miller, R. Time course of neuroleptic therapy for psychosis. Psychopharm. 1987; 92: 405-415.

Mulholland, G., Otto, C., Jewett, D., Kilbourn, M., Koeppe, R., Sherman, P., Petry, N., Carey, J., Atkinson, E., Archer, S., Frey, K., Kuhl, D. Synthesis, rodent biodistribution dosimetry, metabolism and monkey images of Carbon-11-labeled (+)-2 alpha-tropanyl benzilate: a central muscarinic receptor imaging agent. J Nucl Med. 1992; 33: 423-430.

Nordstrom, A-L, Farde L, Halldin C et al. Time course of D2-dopamine receptor occupancy examined by PET after single oral doses of haloperidol. Psychopharm 1992; 106:433-438.

Palacios, J. Distribution of serotonin receptors. Ann New York Acad Sci. 1990; 600: 36-52.

Pazos, A., Probst, A., Palacios, J. Serotonin receptors in the human brain-IV autoradiographic mapping of serotonin 2 receptors. Neurosci. 1987; 21: 123-139.

Peroutka, S. 5-hydroxytryptamine receptor subtypes: molecular, biochemical and physiological characterization. Trends Neurosci. 1989; 11: 496-500.

Perry, T., Buchanan, J., Kish, S., Hansen, S. GABA deficiency in the brains of schizophrenic patients. Lancet. 1979; 1: 237.

Pickar, D. Perspectives on a time-dependent model of neuroleptic action. Schiz Bull. 1988; 14:255-268.

Seeman, P., Guan, H., Niznik, H. Endogenous dopamine lowers the dopamine D2 receptor density as measured by [^{3}H]Raclopride. Synapse. 1989; 3:96-97.

Smith, G., Dewey, S., Logan, J., Brodie, J. Vitkun, S. Simkowitz, P. Alexoff, D. Fowler, J., Volkow, N., Wolf, A. The serotonin-dopamine interaction measured with positron emission tomography (PET) and C-11 raclopride in normal human subjects. J Nucl Med. 1994; 35: 85.

Smith, M., Wolf, A., Brodie, J., Arnett, C., Barouche, F., Shuie, C., Fowler, J., Russel, J., Macgregor, R., Wolkin, A., Angrist, B., Rotrosen, J., Peselow, E. Serial ^{18}F-N-methylspiroperidol PET studies to measure change in antipsychotic D2 drug receptor occupancy in schizophrenic patients. Biol Psychiat. 1988; 23: 653-663.

Smith, Y., Bolam, J. The output neurons and the dopaminergic neurons of the substantia nigra receive a GABA containing input from the globus pallidus in the rat. J Comp Neurol. 1990; 296: 47-64.

Stahl, S., Thornton, J., Simpson, M., Berger, P., Napoliello, M., Gamma-vinyl-GABA treatment of tardive dyskinesia and other movement disorders. Biol Psychiat. 1985; 20: 888-893.

Steinbusch, H. Distribution of serotonin immunoreactivity in the central nervous system of the rat. Neurosci. 1981; 6: 557-618.

Stevens, J. An anatomy of schizophrenia? Arch Gen Psychiat. 1973; 29: 177-189.

Tamminga, C., Thaker, G., Buchanan, R., Kirkpatrick, B., Alphs, L., Chase, T., Carpenter, W. Limbic system abnormalities identified in schizophrenia using positron emission tomography with fluorodeoxyglucose and neocortical alterations with deficit syndrome. Arch Gen Psychiat.; 49: 522-530.

Tamminga, C., Thaker, G., Hare, T. Ferraro, T. GABA agonist therapy improves tardive dyskinesia. Lancet. 1983; II: 97-98.

Tan, P., Dewey, S., Gatley, S., Pappas et al., Drug pharmacokinetics and pharmacodynamics: PET and microdialysis studies of SR 46349B, a selective serotonin 5HT-2 antagonist. J Nucl Med. 1994; 35: 67.

Tandon, R., Greden, J. Cholinergic hyperactivity and negative schizophrenic symptoms. A model of cholinergic/dopaminergic interactions in schizophrenia. Arch Gen Psychiat. 1989; 34: 236-239.

Thaker, G., Tamminga, C., Alpha, L., Lafferman, J., Ferraro, T., Hare, T. Brain gamma aminobutyric acid abnormality in tardive dyskinesia. Arch Gen Psychiat. 1987; 44: 522-529.

Torrey, E., Peterson, M. Schizophrenia and the limbic system. Lancet. 1974; II: 942-946.

Volkow, ND, Brodie JD, Wolf AP, Gomez-Mont F, Cancro R, Van Gelder P, Russell JAG and Overall J: Brain Organization in Schizophrenia. J Cereb Blood Flow Metab 1986; 6:441-446.

Volkow, N., Fowler, J., Wang, G., Dewey, S., Schleyer, D., et al. Reproducibility of repeated measures of Carbon-11 raclopride binding in the human brain. J Nucl Med. 1993; 34: 609-613.

Volkow, N., Wang, G-J., Fowler, J. et. al., Imaging dopamine release in the human brain. Synapse. 1994; 16: 255-262.

Weinberger, D. A connectionist approach to the prefrontal cortex. J Neuropsych Clin Neurosci. 1993; 5: 241-253.

Wolkin, A., Brodie, J., Barouche, F., Rotrosen, J., Wolf, A., Cooper, T. Dopamine receptor occupancy and haloperidol plasma levels. Arch Gen Psychiat. 1989; 46:482-486.

Wolkin, A., Barouche, F., Wolf, A., Rotrosen, J., Fowler, J., Shuie, C., Cooper, T., Brodie, J. Dopamine blockade and clinical response: evidence for two biological subgroups of schizophrenia. Am J Psychiat. 1989; 146: 905-908.

Wolkin, A., Jaeger, J., Brodie, J., Wolf, A., Fowler, J., Rotrosen, J., Gomez-Mont, F., Cancro, R. (1985) Persistence of cerebral metabolic abnormalities in chronic schizophrenia as determined by Positron Emission Tomography. Am J Psychiat. 1985; 145;251-253.

Wolkowitz, O., Pickar, D. Benzodiazepines in the treatment of schizophrenia: a review and reappraisal. Am J Psychiat., 1991; 148: 714-726.

Wong, D., Gjedde, A., Wagner, H., et al. Quantification of neuroreceptors in the living human brain. II Inhibition studies of receptor density and affinity. J Cereb Blood Flow Metab. 1986;6:147-153.

Wong,DF, Wagner HN, Dannals RF et al. Effects of age on dopamine and serotonin receptors measured by positron tomography in the living human brain. Science. 1984; 226: 1393-1395.

Wong, D., Wagner, H., Tune, L., et al. Positron Emission Tomography reveals elevated D2 dopamine receptors in drug naive schizophrenics. Science. 1986; 234:1558-1560.

Zaborszky, L. Afferent connections of the forebrain cholinergic projection neurons. In Central Cholinergic Synaptic Transmission. Frotscher, M., Misgeld, U., (eds.), Basel, Birkhauser, 1991; pp. 12-32.

Zezula, J., Cortes, R., Probst, A., Palacios, J. Benzodiazepine receptor sites in the human brain: autoradiographic mapping. Neurosci. 1988; 22: 771-795.

INDEX